Industrial Engineering Applications in Emerging Countries

Industrial Engineering: Management, Tools, and Applications

Industrial Engineering Non-Traditional Applications in International Settings

Industrial Engineering Applications in Emerging Countries

Global Logistics Management

Industrial Engineering Applications in Emerging Countries

Edited by
İhsan Sabuncuoğlu
Bahar Y. Kara
Bopaya Bidanda

CRC Press is an imprint of the
Taylor & Francis Group, an **informa** business

CRC Press
Taylor & Francis Group
6000 Broken Sound Parkway NW, Suite 300
Boca Raton, FL 33487-2742

International Standard Book Number-13: 978-1-4822-2689-8 (Hardback)

Library of Congress Cataloging-in-Publication Data

Industrial engineering applications in emerging countries / edited by Ihsan Sabuncuoglu, Bahar Y. Kara, and Bopaya Bidanda.
pages cm -- (Industrial engineering)
Includes bibliographical references and index.
ISBN 978-1-4822-2689-8 (hardback)
1. Industrial engineering--Developing countries. I. Sabuncuoglu, Ihsan (Engineer) II. Kara, Bahar Y. (Bahar Yetis) III. Bidanda, Bopaya.

T56.I435 2015
658.5--dc23 2014026732

Visit the Taylor & Francis Web site at
http://www.taylorandfrancis.com

and the CRC Press Web site at
http://www.crcpress.com

This book is dedicated to my wife, Semra, for her kindness and devotion throughout my career and to my parents, Ferhan and Vehbi, for their endless support since my childhood.

İhsan Sabuncuoğlu

Contents

Preface

We are pleased to present to you this book that focuses on recent industrial engineering (IE) applications on selected logistics problems in emerging countries. Even though IE was born and evolved in the United States, the profession has witnessed a rapid growth in the breadth and depth of applications in many other countries in the last three decades. This book provides a collation of modern applications of IE tools and techniques to a diversity of environments in the context of emerging countries.

We therefore believe that this book can play an important role in creating an awareness of the breadth of the IE discipline in terms of both geography and scope.

Editors

İhsan Sabuncuoğlu is the founding rector of Abdullah Gul University. He earned his BS and MS in industrial engineering from the Middle East Technical University in 1982 and 1984, respectively. He earned his PhD in industrial engineering from Wichita State University in 1990.

Dr. Sabuncuoğlu worked for Boeing, Pizza Hut, and the National Institute of Heath in the United States during his PhD studies. He joined Bilkent University in 1990 and worked as a full-time faculty member until 2013. In the meantime, he held visiting positions at Carnegie Mellon University in the United States and at Institut Français de Mécanique Avancée (IFMA) in France. His research interests are in real-time scheduling, simulation optimization, and applications of quantitative methods to cancer-related health-care problems. His research has been funded by TUBITAK (The Scientific and Technological Research Council of Turkey) and EUREKA (a European-wide initiative to foster European competitiveness through cooperation among companies and research institutions in the field of advanced technologies).

Dr. Sabuncuoğlu also has significant industrial experience in aerospace, automotive, and military-based defense systems. His industrial projects are sponsored by a number of both national and international companies. He is currently the director of the Bilkent University Industry and the University Collaboration Center (USIM) and the chair of the Advanced Machinery and Manufacturing Group (MAKITEG) at TUBITAK.

In addition to publishing more than a hundred papers in international journals and conference proceedings, Dr. Sabuncuoğlu has edited two books. He is also on the editorial board of a number of scientific journals in the areas of industrial engineering and operations research. He is also a member of the Institute of Industrial Engineering, the Institute for Operations Research, the Management Sciences, and the Simulation Society. He is also a member of the Council of Industrial Engineering Academic Department Heads (CIEADH) and various other professional and social committees.

Bahar Y. Kara is an associate professor in the Department of Industrial Engineering at Bilkent University.

Dr. Kara earned an MS and a PhD from the Bilkent University Industrial Engineering Department, and she worked as a postdoctoral researcher at McGill University in Canada.

Dr. Kara was awarded Research Excellence in PhD Studies by INFORMS (Institute for Operations Research and Management Science) UPS-SOLA.

In 2008, Dr. Kara was awarded the TUBA-GEBIP (National Young Researchers Career Development Grant) Award. She attended the World Economic Forum in China in 2009. For her research and projects, the IAP (Inter Academy Panel) and the TWAS (The Academy of Science for the Developing World) Awarded her the IAPs Young Researchers Grant. Dr. Kara was elected as an associate member of Turkish Academy of Sciences in 2012. She has been acting as a reviewer for the top research journals within her field. Her current research interests include distribution logistics, humanitarian logistics, hub location and hub network design, and hazardous material logistics.

Bopaya Bidanda is currently the Ernest E. Roth professor and chairman in the Department of Industrial Engineering at the University of Pittsburgh. His research focuses on manufacturing systems, reverse engineering, product development, and project management. He has published five books and more than a hundred papers in international journals and conference proceedings. His recent (edited) books include those published by Springer—*Virtual Prototyping & Bio-Manufacturing in Medical Applications* and *Bio-Materials and Prototyping Applications in Medicine*. He has also given invited and keynote speeches in Asia, South America, Africa, and Europe. He also helped initiate and institutionalize the engineering program on the Semester at Sea voyage in 2004.

He previously served as the president of the Council of Industrial Engineering Academic Department Heads (CIEADH) and also on the board of trustees of the Institute of Industrial Engineers. He also serves on the international advisory boards of universities in India and South America.

Dr. Bidanda is a fellow of the Institute of Industrial Engineers and is currently a commissioner with the Engineering Accreditation Commission of ABET. In 2004, he was appointed a Fulbright Senior Specialist by the J. William Fulbright Foreign Scholarship Board and the U.S. Department of State. He received the 2012 John Imhoff Award for Global Excellence in Industrial Engineering given by the American Society for Engineering Education. He also received the International Federation of Engineering Education Societies (IFEES) 2012 Award for Global Excellence in Engineering Education in Buenos Aires and also the 2013 Albert G. Holzman Distinguished Educator Award given by the Institute of Industrial Engineers. In recognition of his services to the engineering discipline, the medical community, and the University of Pittsburgh, he was honored with the 2014 Chancellors Distinguished Public Service Award.

Contributors

Noor Ashour
Department of Industrial Engineering
American University of the Middle East
Kuwait City, Kuwait

Shaikha Al-Dabbous
Department of Industrial Engineering
American University of the Middle East
Kuwait City, Kuwait

Reem Al-Mady
Department of Industrial Engineering
American University of the Middle East
Kuwait City, Kuwait

Noriah Al-Mas
Department of Industrial Engineering
American University of the Middle East
Kuwait City, Kuwait

Suad Al-Suwaiti
Department of Industrial Engineering
American University of the Middle East
Kuwait City, Kuwait

M.M. Haris Aslam
Department of Operations and Supply Chain
University of Management and Technology
Lahore, Pakistan

Sarwar M. Azhar
School of Business and Economics
University of Management and Technology
Lahore, Pakistan

Tashfeen M. Azhar
School of Business and Economics
University of Management and Technology
Lahore, Pakistan

Fatih Baytugan
Information Technologies
Turkish Electricity Transmission Company
Ankara, Turkey

Selim Bora
Department of Mechanical Engineering
Texas A&M University at Qatar
Doha, Qatar

Kürşad Derinkuyu
Faculty of Business Administration
Department of Logistics Management
University of Turkish Aeronautical Association
Ankara, Turkey

Banu Y. Ekren
Department of Industrial Engineering
Izmir University of Economics
Izmir, Turkey

Mahmut Ekşioğlu
Ergonomics Laboratory
Department of Industrial Engineering
Boğaziçi University
İstanbul, Turkey

Shaima Goodarzi
Department of Industrial Engineering
American University of the Middle East
Kuwait City, Kuwait

Rawan Jaragh
Department of Industrial Engineering
American University of the Middle East
Kuwait City, Kuwait

Chibli Joumaa
Department of Industrial Engineering
American University of the Middle East
Kuwait City, Kuwait

Seifedine Kadry
Department of Industrial Engineering
American University of the Middle East
Kuwait City, Kuwait

Gino J. Lim
Department of Industrial Engineering
University of Houston
Houston, Texas

N.S. Narahari
Department of Industrial Engineering and Management
R.V. College of Engineering
Bangalore, India

Adalet Oner
Department of Industrial Engineering
Yasar University
Izmir, Turkey

M. Arslan Ornek
Department of Industrial Engineering
Izmir University of Economics
Izmir, Turkey

Guler Ozturk
Department of Industrial Engineering
Yasar University
Izmir, Turkey

Hamid R. Parsaei
Department of Mechanical Engineering
Texas A&M University at Qatar
Doha, Qatar

V. Prashant
Department of Industrial Engineering and Management
R.V. College of Engineering
Bangalore, India

Abdul Raouf
University of Management and Technology
Lahore, Pakistan

Mustafa Sezgin
Electricity Market Division
Energy Market Regulatory Authority
Ankara, Turkey

Shahad Sheer
Department of Industrial Engineering
American University of the Middle East
Kuwait City, Kuwait

K.N. Subramanya
Department of Industrial Engineering and Management
R.V. College of Engineering
Bangalore, India

Fehmi Tanrısever
Faculty of Business Administration
Bilkent University
Ankara, Turkey

M.N. Vijayakumar
Department of Industrial Engineering and Management
R.V. College of Engineering
Bangalore, India

Ijaz Yusuf
School of Business and Economics
University of Management and Technology
Lahore, Pakistan

Introduction

Logistics management is one of the primary topics of industrial engineering. This book is the second volume of a three-volume series on global logistics management and focuses on recent developments and implementations of operations research techniques on selected logistics problems in emerging countries. The book is composed of 10 chapters on topics ranging from quality management in pharmaceutical supply chains to risk analysis of maritime ports.

Chapter 1 focuses on identifying the quality management practices that impact supply chain performance using empirical data from the pharmaceutical manufacturing sector in Pakistan. This chapter provides a global perspective from a developing country's point of view.

As a result of its rapidly developing economy and the growth in the volume of imports and exports, Qatar's ports have experienced ever-increasing traffic in recent years. As many of the vessels are oil or liquefied natural gas (LNG) tankers, the risk of accidents, collision, and terrorist attacks, which may result in leaks or fires, should be managed. Chapter 2 presents a simulation-based maritime risk analysis study.

Chapter 3 focuses on Turkish day-ahead electricity markets. Turkish electricity markets have recently evolved into a complex competitive business environment with an increasing role of the private sector in production, consumption, and retailing of electricity. The authors present alternative formulations and solution methods for day-ahead markets' price optimization problem.

In practice, production systems are prone to costly interruptions due to several disruptions such as machine failures. In chemical industries, the lack of required raw materials appears to be an additional cause of disruption, as the usage amount of some types of chemicals (such as fillers) may not be accurately estimated in advance. Because interruptions are costly, chemical industries tend to store such chemicals as floor stock items at periphery of the production areas for use when needed. Chapter 4 presents a floor stock inventory control model that uses simulation optimization.

Human muscle is still the main source of power in many industrial tasks. Chapter 5 presents a critical review of hand torque strength studies with the aim of providing a perspective for better application of torque strength norms to control the risks of musculoskeletal disorders, increase the productivity and quality of work in the workplace, and design safe and usable consumer products.

Another important problem encountered in emerging countries is traffic congestion due to a developing economy and, hence, an ever-increasing number of vehicles. Chapter 6 analyzes the traffic flow on Kuwait's roads and highways and proposes remedies to increased congestion levels.

Even though renewable energy sources have gained an increasing interest in the recent years, fossil fuels and especially petroleum continue to get the lion's share in many countries. Chapter 7 models and analyzes the production line of a crude oil refinery in Kuwait via simulation.

Another transportation application is presented in Chapter 8. The idea that public transportation relieves road congestion is popular. The authors analyze the effectiveness of Izmir's (the third largest city in Turkey) subway network via simulation.

Chapter 9 focuses on productivity measurement and improvement in process industries. After a brief review, the authors demonstrate the effects of several improvement ideas on a biscuit manufacturing system.

Chapter 10 presents another interesting case study. The authors analyze the operations of a Pakistani potato-based starch producer via a system dynamics model with the goal of optimizing inventory levels.

Special thanks to Dr. Selcuk Goren from AGU for his great effort and help in editing this book.

1

Quality Management in Pharmaceutical Supply Chains

Developing-Country Perspective

M.M. HARIS ASLAM AND ABDUL RAOUF

Contents

1.1 Introduction

Quality management (QM) is a philosophy that targets continuous improvement of all the business practices and processes with the focus on customer satisfaction, while involving all the internal and external stakeholders of the organization. It has received significant attention in the operations management research over the years (Nair, 2006). Effective QM is attributed to the achievement of results such as improved external and internal focus, better design, strengthening of weak processes, and protection of strong areas (Fotopoulos and Psomas, 2010). There is considerable empirical evidence relating QM to organizational success. A large number of empirical studies have shown that QM is positively related to organizational performance (Awan et al., 2009; Das et al., 2008; Flynn et al., 1995; Fotopoulos and Psomas, 2010; Kaynak, 2003; Mohrman et al., 1995; Prajogo and Sohal, 2006; Samson and Terziovski, 1999) and competitive advantage (Douglas and Judge, 2001; Powell, 1995).

Recent years have seen proliferation of studies considering QM in supply chain (SC) perspective (Choi and Rungtusanatham, 1999; Flynn and Flynn, 2005; Foster et al., 2010; Kannan and Tan, 2007; Kaynak and Hartley, 2008; Kuei and Madu, 2001; Kuei et al., 2010; Lin et al., 2005; Ramos et al., 2007; Sila et al., 2006; Zhang et al., 2011). This is a result of realization by the organizations that getting the product to the customers is not enough and there is a need to be more responsive to the changing customer requirements (Kannan and Tan, 2007). There has also been an increased reliance upon outsourced materials and processing coupled with reduced supplier base and greater information sharing with customers and suppliers (Chandra and Kumar, 2000). SC perspective accentuates the interdependence of buyer–supplier collaboration to improve the performance of the whole SC. According to supply chain management (SCM) approach, companies do not seek to maximize their profitability at the expense of their SC partners (Romano and Vinelli, 2001). This is in line with the QM philosophy where process optimization as opposed to activity optimization is emphasized. Extension of QM practices in SCs is thus required since the quality delivered to the end customer is the outcome of value added by each entity in the SC in terms of product/service features, cost reduction, time, and flexibility to customer demands.

In order to evaluate SC-wide improvements, there is a need for designing measures for supply chain performance (SCP). Complexity of SCs makes the task of designing such measures very challenging (Beamon, 1999). Literature on SCP consists of two types of studies: studies identifying performance measures for the SCs and studies identifying factors affecting the SCP (Sezen, 2008). This chapter focuses on the latter aspect.

Sila and Ebrahimpour (2003) in a meta-analysis of QM studies showed that there is a dearth of information about the nature of QM implementation in Asia, South America, Africa, and the Middle East. They emphasized the need for studies to be conducted in different countries in order to identify the critical QM practices in the context of these particular countries. Research on various aspects of QM in Pakistan has not been scarce (e.g., Ahsan, 2010; Awan et al., 2009; Kureshi et al., 2010) yet there are very few studies that have linked QM to performance. Similarly, in an elaborate search for the research studies in the area of SCP, none could be found originating from Pakistan. This chapter is targeted toward filling these gaps. The research question addressed in this chapter is as follows.

1.1.1 Does the QM in SCs Impact the Performance of the SCs?

This research question has been addressed in the context of pharmaceutical industry of Pakistan. Pharmaceutical products are an important part of the health-care system of a country (Obaid, 2009). The importance of product quality to this industry cannot be overstated. SC effectiveness in this industry is a challenge; however, achieving it brings high rewards (Asamoah et al., 2011). Pharmaceutical industry of Pakistan consists of over 400 national and 25 multinational companies (Pakistan Pharmaceutical Manufacturers Association, 2011). The market share is split equally between national and multinational companies (Epsicom, 2011). Industry worth exceeds $1.4 billion, provides employment to over 220,000 people directly or indirectly, and earns a foreign exchange of over 100 million every year (Shahrukh, 2011). The local industry is able to meet 70% demand for the finished medicine (Zaheer, 2011). The analysis has been performed from the perspective of the focal firms. Focal firms for this study are pharmaceutical manufacturing organizations. In the context of this study, supplier–focal firm dyad involves flow of raw materials, funds, and information. Most of the suppliers are located outside the country (Epsicom, 2011). Focal firm–customer dyad includes finished products, funds, and information. Three types of customers exist for the pharmaceuticals manufacturers: distributors, medical institutions (hospitals), and retail chains.

The remaining part of this chapter is organized as follows. The next section presents a review of literature on QM and SCP, followed by methods, results, and discussion of results. The conclusion section discusses the contribution of this research and provides suggestions for future research.

1.2 Literature Review

1.2.1 Quality Management in Supply Chains

QM is a philosophy that focuses on continuous improvement of all organizational functions. This improvement is only achievable if quality concepts are employed from raw material acquisition to after sales customer service (Kaynak, 2003). While traditionally the onus of quality improvement activities was on the quality departments of the organizations, it is being realized that it is everyone's responsibility, from the suppliers of the raw materials to the distributors of the finished goods (Forker et al., 1997). This has resulted in an increased interest in the area of QM in the context of SCs during the recent years (Lin et al., 2005). SC quality strategy and competence thus are gaining importance (Kuei et al., 2008) in helping organizations attain market responsiveness with precision and profitability (Kuei et al., 2001). SC integration necessitates process quality throughout the SC, which can result in cost reduction, higher utilization of resources, and higher efficiency (Beamon and Ware, 1998). In pursuit of achieving product and process quality, many benefits can accumulate from managing the flow of materials and information in the end-to-end SC (Romano and Vinelli, 2001). Traditionally, company-centric issues like price, product quality, and delivery time were considered important by the companies, whereas in QM from an SC perspective, supplier–customer relationships, combined effort on production of quality products, etc., have become central issues (Lin et al., 2005). The evolution of quality from internal focus to the current world-class quality with the idea of joint quality policy making by SC partners can be depicted in Figure 1.1.

A host of studies have appeared in the last decade or so, which have researched the area of QM in SCs (e.g., Das, 2011; Flynn and Flynn, 2005; Foster, 2008; Foster et al., 2010; Fynes et al., 2005; Kannan and Tan, 2007; Kaynak and Hartley, 2008; Kuei et al., 2001, 2008, 2010; Robinson and Malhotra, 2005; Sila et al., 2006;

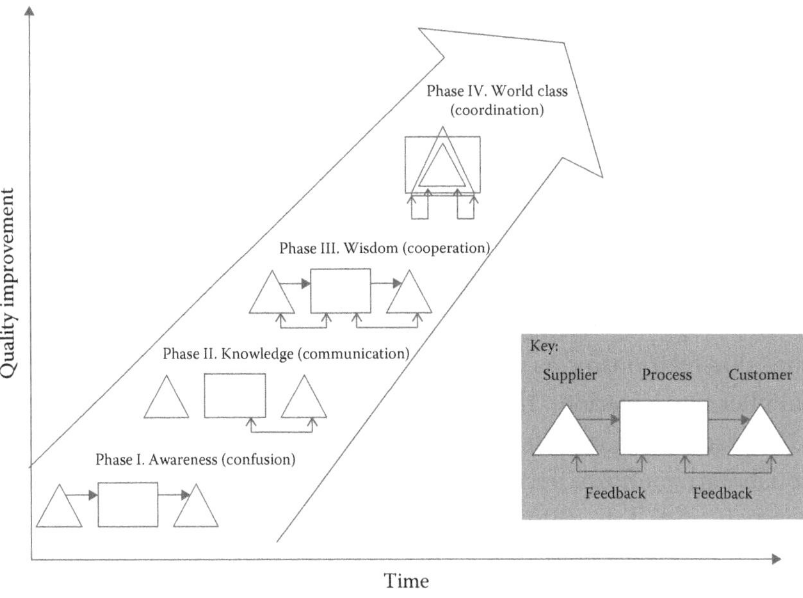

Figure 1.1 QM evolution. (From Raouf, A. and Yusaf, I., *Quality Management: A Customer Driven Approach*, University of Management and Technology Press, Lahore, Pakistan, 2011.)

Vanichchinchai and Igel, 2010). One of the earlier studies in the field was conducted by Forker et al. (1997). They studied the relationship between QM practices and performance in an electronics SC. The empirical findings of this study suggested a significant relationship between supplier quality across the SC and quality performance. Based on empirical findings, the authors suggested that the role of quality departments in the organizations should be improved; quality from suppliers as well as suppliers' suppliers should be emphasized; quality data should be collected in a timely manner and made available to all the employees; and the role of employees in the quality improvement efforts should be enhanced through mechanisms such as training, feedback, and rewards and recognition. Kuei et al. (2001) argued that in order to sustain the customer-driven culture, and to offer the right products at the right place and at the right time, organizations need to nurture a quality culture for improving customer satisfaction, employee satisfaction, and organizational performance throughout the SC. Sila et al. (2006) examined various aspects of quality in SCs using survey methodology. The results of their study showed that quality in SCs is significantly related to the quality of the final product. Lai et al. (2005) explored the effect of relationship stability between the buyer and the supplier firm on the supplier commitment to quality. Results of their study showed that stable relationships are positively related to supplier commitment to quality for the buyer firm where suppliers' perception of supply certainty makes this relationship stronger, while asset specificity and transaction frequency do not have any impact on this relationship. Lin et al. (2005) in their empirical investigation of relationship between QM and organizational performance showed that this relationship is mediated by supplier selection strategy and supplier participation. This led to the conclusion that incorporating QM practices in supplier participation programs could support the collaborative efforts necessary for improved organizational results. Kannan and Tan (2007) identified customer input and supplier quality as *externally focused* quality practices that show firms' orientation toward managing SC quality. The empirical findings of their study led to the conclusion that the externally focused quality practices have a stronger impact on the performance than the internally focused practices. Kuei et al. (2008) presented a strategic framework based on four drivers of SC quality, namely, SC competence, critical success factors, strategic components, and SC quality practices that can help multinational organizations in implementing SC QM plans. Kaynak and Hartley (2008) pointed out that SCM consists of internal practices of the firm and external practices that cut through the organizational boundaries to integrate the organization with its customers and suppliers. Collaboration with customer and supplier is also a goal in QM; hence, supplier QM (SQM) and customer focus are two quality practices that extend QM into the SC. The results of their study showed the interrelationships between quality practices and their significant relationship with various aspects of organizational performance. Soltani et al. (2011) emphasized the importance of role played by manufacturers in ensuring product quality in the global SCs.

In summary, achieving increased customer focus throughout the SC and managing supplier relations in a mutually beneficial manner are goals in both SCM and QM. However, QM encompasses many more aspects than these two practices. This chapter explores how the various quality practices impact the SCP.

1.2.2 *Supply Chain Performance*

Lately, the focus of performance measurement (PM) research has shifted to PM of the whole SC as opposed to individual companies. The realization that organizational success is dependent upon the success of the whole SC is a contributing factor to this shift (Zelbst et al., 2009). PM in the SCs helps management in designing actions targeted toward achieving performance improvement and increased competitiveness (Van Hoek, 1998). SCP measures differ from traditional measures in that they do not rely heavily on the tangible financial measures like profit and return on investment (Saad and Patel, 2006), which are typically internally focused (Lambert and Pohlen, 2001) and backward looking in nature. Lambert and Pohlen (2001) suggested that the probability of success of an SC is significantly increased by a well-designed PM system. Chan (2003) suggested that SCs need to be managed as one company and hence the performance measures are required that represent the performance of the whole SC as opposed to any single entity. Gunasekaran et al. (2001) also emphasized the need for PM that can lead to fully integrated SCs. The process of designing measures of SCP is complex due to the complexity of SCs (Lambert and Pohlen, 2001). Various frameworks have been prescribed in the literature for measuring the SCP. Some notable frameworks have been summarized in Table 1.1.

Based on these frameworks, it can be deduced that SCP is a function of the ability of an organization to provide the right product

Table 1.1 Frameworks for SCP Measurement

STUDY	FRAMEWORK FOR MEASURING SCP
Sezen (2008)	Flexibility, output, and resource performance
Lee et al. (2007)	Cost containment and reliability
Forslund and Jonsson (2007)	Corrective action, preventive action, and customer service performance
Hausman (2004)	Service, assets, and speed
Chan (2003)	Cost, time, capacity, capability, productivity, utilization, and outcome
Agarwal and Shankar (2002)	Market sensitiveness, process driven, and process integration
Brewer and Speh (2000)	End customer benefits, SCM improvement, SCM goals, and financial benefits
Beamon (1999)	Flexibility, output, and resource performance

(responsiveness to customer needs in terms of product variety and demand variation), at the right place (order delivery precision), at the right time (time-based performance), and at the right price (cost-based performance).

1.2.3 Conceptual Framework

The impact of QM on organizational performance has long been advocated by the quality gurus and various researchers in the field. Various researchers have studied the relationship between QM in SCs and different indicators of organizational performance (e.g., Kannan and Tan, 2007; Kaynak and Hartley, 2008; Kuei et al., 2001; Lin et al., 2005; Ou et al., 2010; Sila et al., 2006; Tan et al., 1999). Tan et al. (1999) showed that effective integration of suppliers, manufacturers, and customers is directly related to the achievement of financial and growth objectives. On the basis of a survey of ISO 9000 certified companies, Casadesús and de Castro (2005) showed that quality assurance through QM systems resulted in improvement in relationships with suppliers and customer satisfaction. Yeung (2008) argued that QM provides the basis for strategic supply management and is important for improvement throughout the SC. In order to minimize overall costs, lead time, and transportation time and improve customer service levels in the SCs, quality and technology-driven capabilities have to be maintained (Kuei et al., 2002). Flynn and Flynn (2005) argued that QM results in reduction of safety stock, rework, cycle times, pipeline inventory, response time, inspection, and cycle stock requirements. The result of their survey showed that goals of QM and SCM are pursued simultaneously and QM significantly impacts SCP. Vanichchinchai and Igel (2009) argued that both QM and SCM have same ultimate goal of customer satisfaction. In their study on Thai automotive industry, they found a significant relationship between QM and firm's supply performance (Vanichchinchai and Igel, 2010). Current research studies the relationship between QM in the SCs and SCP. Conceptual framework for the chapter is provided in Figure 1.2.

The following section narrates the methods employed in this chapter followed by the section providing the analysis procedures.

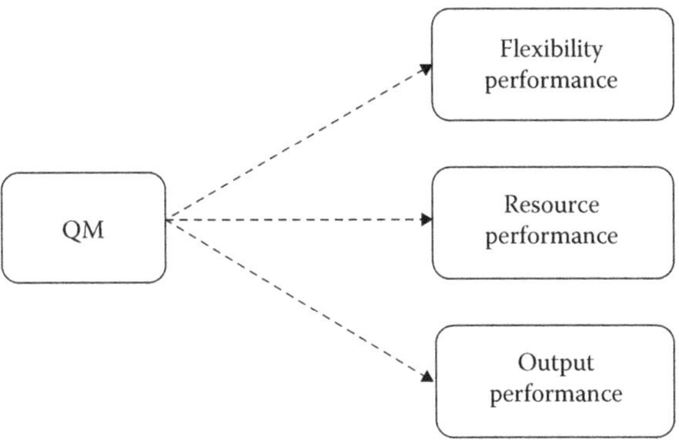

Figure 1.2 Conceptual model for the study.

1.3 Methods

1.3.1 Survey Instrument

Measures for the constructs related to QM were adopted from Kaynak and Hartley (2008). This framework consists of eight quality practices, namely, management leadership, training, employee relations, customer focus, quality data and reporting, SQM, product/service design, and process management. Measures related to SCP were adopted from Beamon (1999) consisting of three SCP indicators, namely, flexibility performance, resource performance, and output performance. Final survey instrument consisted of two sections: section 1 consisted of 40 items related to each of the 8 quality practices identified in the literature review and section 2 contained 17 items related the 3 SCP indicators. Five-point Likert scales were used for the measurement of all the constructs. Responses for the quality constructs ranged between *very low* (=1) to *very high* (=5). Regarding the SCP constructs, respondents were requested to judge their organizations' performance as compared to major industry competitors on the scale of *worse* (=1) to *better* (=5).

The instrument was pretested before administration to the target respondents. Twenty respondents from the pharmaceutical manufacturing industry were selected for the pilot test. Both quantitative and qualitative responses were taken from these respondents regarding the survey instrument. Quantitative analysis was limited to computation of Cronbach's alpha (Kaynak, 2003; Kaynak and Hartley, 2008).

Alpha values for all QM and SCP constructs were above 0.7 except for SC output performance (0.69). Based on the qualitative judgments of professionals in pharmaceutical sector, one item from the customer focus scale and two items from the SQM scale were eliminated from the instrument. The final instrument was reviewed by senior faculty at the *department of operations and supply chain* at a leading national university before sending to target respondents.

1.3.2 Sampling and Survey Procedures

This study employed survey method, which is the most frequently used methodology for data collection in areas that lack sufficient secondary data (Kannan and Tan, 2007). Target respondents were identified as managers working in the SC, logistics, procurement, distribution, or quality department. A cover letter containing the link to survey was e-mailed to 200 managers working in aforementioned capacities in pharmaceutical manufacturing companies. Initial mailing did not generate any noteworthy response. Previous studies have shown that Asian firms show reluctance in cooperation if no prior relationship exists (Carr et al., 2000). Similar problems were also faced by Vanichchinchai and Igel (2010) while collecting data in Thai automotive manufacturing industry. In the light of disappointing response to the first round of e-mails, snowballing technique was used in order to reach the target respondents in the second round. This effort generated 73 responses of which 68 were considered valid for the analysis after the initial screening. Table 1.2 shows the description of the sample.

1.4 Results

1.4.1 Measure Validation

There are three types of validity: content validity, construct validity, and criterion-related validity. Content validity (or face validity) subjectively appraises the correspondence between individual items and the concepts that are measured through these items (Hair et al., 2007). It is established among other means through literature review, ratings by expert judges, and pilot testing. Eight constructs for QM and three constructs for SCP have content validity since they were

Table 1.2 Sample Description

	FREQUENCY	%
ORGANIZATION TYPE		
National	38	55.9
Multinational	27	39.7
N/A	3	4.4
DEPARTMENT		
Distribution	28	41.18
Logistics	4	5.88
Production	5	7.35
Quality	13	19.12
Supply Chain	16	23.53
Others	2	2.94
APPROXIMATE REVENUE		
Less than 250 million	4	5.9
250–500 million	8	11.8
501–750 million	3	4.4
751–1000 million	10	14.7
Above 1000 million	22	32.4
N/A	21	30.9
Total	68	100

identified after an extensive review of literature. Evaluation by experts both in the field as well as academics further substantiates the content validity of this instrument. Criterion-related validity (or predictive validity) is the extent to which a model is related to an independent measure of related criterion (Samson and Terziovski, 1999). Criterion-related validity was the degree to which quality practices in SCs were related to SCP measures. In order to assess the criterion-related validity, bivariate correlations were computed between the dependent and independent variables. The results are provided in Table 1.3. Bivariate correlation coefficients show that, except for a few exceptions, quality practices are significantly related to SCP measures.

Construct validity is a measure of appropriateness of a set of items to measure a latent variable (Forza and Filippini, 1998). In order to establish construct validity, exploratory factor analysis (EFA) using principal components analysis (PCA) technique was used. EFA is applied to a set of variables to identify which subsets of the variables are related to each other but are different from others. These variables

Table 1.3 Correlation Matrix

	ML	T	ER	CF	QDR	SQM	PD	PM	FP	RP	OP
ML	1	0.569**	0.462**	0.604**	0.568**	0.540**	0.527**	0.466**	0.544**	0.397**	0.537**
T			0.517**	0.446**	0.521**	0.371**	0.423**	0.342**	0.510**	0.165	0.339**
ER				0.447**	0.528**	0.306*	0.241*	0.353**	0.396**	0.193	0.308*
CF					0.444**	0.518**	0.428**	0.532**	0.529**	0.291*	0.428**
QDR						0.596**	0.522**	0.451**	0.435**	0.300*	0.577**
SQM							0.559**	0.505**	0.472**	0.136	0.381**
PD								0.476**	0.279*	0.355**	0.377**
PM									0.469**	0.296*	0.413**
FP										0.172	0.540**
RP											0.483**
OP											1

Note: ML, management leadership; T, training; ER, employee relations; CF, customer focus; QDR, quality data and reporting; SQM, supplier quality management; PD, product design; PM, process management; FP, financial performance; RP, resource performance; OP, output performance.

*Correlation is significant at the 0.05 level (two tailed).

**Correlation is significant at the 0.01 level (two tailed).

are considered to be representatives of the processes that have created these correlations (Tabachnick et al., 2007). In order to improve the ease of interpretation, varimax rotation was employed. Separate analyses were performed for dependent and independent variables due to the small sample size. Bartlett's test of sphericity (significance value <0.05) and the Kaiser–Meyer–Olkin measure of sampling adequacy (>0.70) were used to validate the use of factor analysis. Initially, 37 and 17 items were entered into the analysis. Measures with loading of 0.5 or more were retained. After eliminating the items with insignificant loading and significant cross loadings on multiple factors, 26 and 13 items were retained for QM and SCP scales, respectively. None of the factors was eliminated.

A measure is termed reliable if it is not affected by random error (Forza and Filippini, 1998). A widely used measure of reliability is internal consistency measured through Cronbach's alpha coefficient (Cronbach, 1951). Scales validated through PCA were then used to compute the alpha coefficients and summated scores. Final output of factor analysis is shown in Tables 1.4 and 1.5 along with alpha coefficients, means, and standard deviations for the scales.

1.4.2 Regression Analysis

Multiple linear regression was used in order to study the impact of QM on SCP. Three separate regression models were constructed since there were three dependent variables. The process was initiated by testing the assumptions regarding linearity, homoscedasticity, normality and independence of the residuals through the residual plots, normal probability plots, and Durbin–Watson statistic. After testing the assumptions, regression models were formulated. Regression equations from these models have been shown in Table 1.6.

The results from the three models show that all three models are statistically significant ($p < 0.05$). Most of the variables in the three models were significant with $p < 0.05$ with the exception of management leadership, SQM, and product design in the model with flexibility performance as the dependent variable, which were significant with $p < 0.10$. The model details have been shown in Tables 1.7 through 1.9. Durbin–Watson statistics for all three models are

Table 1.4 Principal Components Analysis (Independent Variables)

VARIABLES	FACTOR LOADING	CRONBACH'S ALPHA	MEAN	STANDARD DEVIATION
Management leadership (ML)		0.9	3.98	0.47
Degree to which organization top management (top organization executive and major department heads) is evaluated for quality performance	0.820			
Extent to which the organizational top management has objectives for quality performance	0.691			
Degree of participation by major department heads in the quality improvement process	0.622			
Degree to which the organizational top management considers quality improvement as a way to increase profits	0.519			
Amount of review of quality issues in organizational top management meetings	0.517			
Training (T)		0.88	3.42	0.51
Quality-related training given to hourly employees throughout the organization	0.858			
Specific work-skills training (technical and vocational) given to hourly employees throughout the organization	0.743			
Training in statistical techniques within the organization as a whole	0.518			
Employee relations (ER)		0.66	3.60	0.52
Degree of participation in quality decisions by hourly/nonsupervisory employees	0.757			
Extent to which building quality awareness among employees is ongoing	0.523			
Customer focus (CF)		0.9	3.96	0.53
Extent to which managers are aware of the results of customer satisfaction surveys	0.828			
Extent to which managers have access to a summary of customer complaints	0.812			
Extent to which customer satisfaction surveys are used in determining/identifying customers' requirements	0.659			

Quality data and reporting (QDR)		0.85	3.65	0.52
Extent to which quality data (cost of quality, defects, errors, scrap, etc.) are used as tools to manage quality	0.786			
Availability of quality data (error rates, defect rates, scrap, defects, etc.)	0.785			
Timeliness of the quality data	0.653			
Availability of procedures to ensure the reliability and improvement of data gathering	0.535			
Supplier quality management (SQM)		0.63	3.93	0.52
Extent to which long-term relationships are offered to suppliers	0.668			
Thoroughness of your organization's supplier rating system	0.501			
Product design (PD)		0.88	3.52	0.54
Extent to which implementation/producibility is considered in the product/service design process	0.797			
Thoroughness of new product/service design reviews before the product/service is produced and marketed	0.727			
Quality of new products/services emphasized in relation to cost or schedule objectives	0.726			
Process management (PM)		0.81	3.76	0.43
Extent to which process design is *foolproof* and minimizes the chances of employee errors	0.810			
Extent to which employees are authorized to stop production for quality problems	0.646			
Stability of production schedule/work distribution	0.645			
Degree of automation of the process	0.610			

Table 1.5 Principal Components Analysis (Dependent Variables)

VARIABLES	FACTOR LOADING	CRONBACH'S ALPHA	MEAN	STANDARD DEVIATION
Flexibility performance (FP)		0.83	3.82	0.59
Compared to your major competitors, your ability to respond to and accommodate periods of poor supplier performance	0.821			
Compared to your major competitors, your ability to respond to and accommodate periods of poor delivery performance	0.820			
Compared to your major competitors, your ability to respond to and accommodate periods of poor manufacturing performance (machine breakdowns)	0.720			
Resource performance (RP)		0.93	3.58	0.49
Compared to your major competitors, your total cost of resources used	0.913			
Compared to your major competitors, your total cost of distribution, including transportation and handling costs	0.907			
Compared to your major competitors, your costs associated with held inventory	0.882			
Compared to your major competitors, your total cost of manufacturing, including labor, maintenance, and rework costs	0.827			
Compared to your major competitors, your return on investments	0.738			
Output performance (OP)		0.9	3.91	0.42
Compared to your major competitors, your on-time deliveries	0.859			
Compared to your major competitors, your order fill rate	0.809			
Compared to your major competitors, your customer response time	0.768			

Table 1.6 Regression Models

SCP MEASURES	MODEL
Flexibility performance	FP = −0.008 + 0.33 ML + 0.33 T + 0.257 SQM − 0.243 PD + 0.325 PM
Resource performance	RP = 1.94 + 0.415 ML
Output performance	OP = 1.605 + 0.281 ML + 0.334 QDR

Table 1.7 Model Parameters: Flexibility Performance as Dependent Variable

	COEFFICIENTS	T	P-VALUE	VIF
Constant	−0.008	−0.014	0.989	
ML	0.33	1.983	0.052	1.963
T	0.331	2.467	0.016	1.532
SQM	0.257	1.803	0.076	1.77
PD	−0.243	−1.79	0.078	1.727
PM	0.325	2.069	0.043	1.51
Adjusted R^2		0.4030		
p-Value		0.0000		
Durbin–Watson statistic		1.258		

Table 1.8 Model Parameters: Resource Performance as Dependent Variable

	COEFFICIENTS	T	P-VALUE
Constant	1.924	4.065	0
ML	0.415	3.516	0.001
Adjusted R^2		0.1450	
p-Value		0.0010	
Durbin–Watson statistic		1.637	

Table 1.9 Model Parameters: Output Performance as Dependent Variable

	COEFFICIENTS	T	P-VALUE	VIF
Constant	1.605	4.394	0.000	
ML	0.281	2.645	0.010	1.476
QDR	0.324	3.435	0.001	1.476
Adjusted R^2		0.380		
p-Value		0.000		
Durbin–Watson statistic		1.812		

between 1 and 3 indicating the assumption of independence of residuals is met (Field, 2009). Variance inflation factors for the two models with more than 1 significant independent variables were less than 10 indicating the absence of multicollinearity (Tabachnick et al., 2007).

The next section provides the discussion of results.

1.5 Discussion

Results of this research show that QM in pharmaceutical SCs positively impacts the SCP. A comprehensive set of measures related to QM and SCP were used to study this relationship. The data collected from the pharmaceutical manufacturing sector of Pakistan were factor analyzed to identify the critical QM practices and critical measures of SCP in Pakistan's pharmaceutical SCs. Eight QM practices and three SCP indicators were identified. Subsequent regression analysis showed that management leadership, training, SQM, product design, and process management impact the SC flexibility performance. Management leadership impacts SC resource performance, while management leadership and quality data and reporting impact the SC output performance. These results are consistent with the overall results reported by Flynn and Flynn (2005) and Vanichchinchai and Igel (2010), the two studies that have previously explored the relationship between QM and SCP. These results also corroborate the results of the studies that have examined direct or indirect relationships between QM and various elements of organizational performance (Ahire et al., 1996; Kaynak, 2003; Kaynak and Hartley, 2008; Powell, 1995; Prajogo and Sohal, 2003; Sila and Ebrahimpour, 2005; Terziovski and Samson, 1999).

The results emphasize the importance of top management's support for the success of quality initiatives in the organizations. This is consistent with the theory of QM and the prescriptions of quality gurus that suggest top management's commitment to quality plays the most important role in the success of quality programs by influencing the success of other quality practices (Kaynak, 2003). This is also consistent with the studies that have examined the direct and indirect impact of management leadership on the organizational performance (Kaynak, 2003; Kaynak and Hartley, 2008; Nair, 2006;

Sila and Ebrahimpour, 2005). Sila and Ebrahimpour (2003) in their meta-analytic study showed that factors related to top management commitment to quality were extracted in 67 of 73 studies taken from 23 countries, showing the cross-cultural validity of this factor. Leadership helps in the quality improvement efforts by taking initiative in quality culture adoption (Fotopoulos and Psomas, 2010). The responsibility of the top management in the context of SCs extends beyond managing the quality initiatives within the company to promoting participation and providing support for quality initiatives among all channel members (Robinson and Malhotra, 2005). Significant correlations between management leadership and SQM and customer focus (Table 1.3) show the importance of top management's role in managing SC-related quality issues. Besides management leadership, other significant quality practices were training, SQM, quality data and reporting, product design, and process management. Training is an important element of human resource management that impacts customer focus, which in turn impacts the whole SC (Kaynak and Hartley, 2008). Training related to employees' job and basic statistical techniques can increase the motivation for better job performance and equips the employees with tools that allow them to participate more effectively in process improvement activities. SQM was found to be significantly related to flexibility performance. Nair's (2006) results in a meta-analytic study showed that SQM was significantly related to both operational and aggregate organizational performance measures. This result shows the importance of building supplier relationships in order to provide flexibility in the product or service offering to the customer. Organizations that enter into long-term relationships with their suppliers and have thorough supplier ratings systems can get support from their suppliers in terms of both volume-based efficiencies and responsiveness to varying customer demand. Product design even though significant in explaining flexibility performance did not have the expected (positive) sign. This is due to the fact that product design is not an integral part of Pakistani pharmaceutical manufacturing industry since approximately 50% of the market is governed by multinationals (PPMA, 2011) that do not design their products in Pakistan. Process management was also significantly related to flexibility performance. The importance of process management throughout the SCs has been emphasized by various researchers (Beamon and Ware, 1998; Robinson and Malhotra, 2005). Quality data and reporting was found to significantly impact the output performance. Various previous studies have shown that there is indirect relationship between quality data and organizational performance (Kaynak and Hartley, 2008; Nair, 2006; Sila and Ebrahimpour, 2005). After developing internal systems related to quality data collection and analysis, firms need to consider developing systems that involve information and analysis throughout the SC (Srivastava, 2008).

The quality practices that did not impact any indicator of SCP were employee relations and customer focus. Customer focus was the second most extracted factor in Sila and Ebrahimpour's (2003) study. This factor was extracted in 53 studies in 21 countries. These two quality practices also did not impact the SCP significantly in Flynn and Flynn's (2005) study as well. This however is not enough to conclude that these two practices are not related to SCP. Bivariate correlations in Table 1.3 show that these two practices are related to management leadership, which in turn impacts the SCP. Hence, it can be argued that both these practices are indirectly related to the SCP. This substantiates the argument of the researchers who suggest that rather than implementing specific quality tools and practices, complete quality culture is the necessary condition for the success of quality initiatives (Powell, 1995). Significant correlations between the quality practices (see Table 1.3) also suggest that simultaneous implementation of all the quality practices is likely to bear fruit rather than piecemeal approach (Ahire et al., 1996; Tan et al., 1999). It can also be argued on the basis of this finding that simultaneous coordinated implementations of QM practices throughout the SC are likely to be more successful than uncoordinated efforts with internal focus.

Prior to this study, the authors could find no research in Pakistan studying the relationship between QM in SCs and SCP. The study by Awan et al. (2009) is probably one of the very few studies that has covered a broad array of quality practices in relating QM in SCs to performance using pharmaceutical distributors as focal firms. In their study conducted on the pharmaceutical distribution sector, they showed that process design was the most significant quality practice in explaining the performance of pharmaceutical distribution sector in Pakistan. In another similar study on the telecommunication sector, Khan (2010) concluded that QM practices were existent in the sector and were significantly related to each other and hence required a simultaneous application. Awan et al. (2008) in an empirical study also concluded that top management commitment to quality is the most important factor in determining the success or failure of quality initiatives.

1.6 Conclusion and Implications

It is a well-established fact that product and process quality improvement can lead to competitiveness. As the organizations realize that they are reliant upon their SC partners for the success of all their quality improvement efforts (Tan et al., 1999), it has become increasingly important to consider SCP measures to gauge the success of improvement efforts. The objective of this chapter was to identify the quality practices that impact the SCP. Based on empirical data from pharmaceutical manufacturing sector of Pakistan, it was shown that management leadership, training, process management, SQM, and quality data and reporting are the quality practices that are critical to the performance of pharmaceutical SCs. It can also be concluded that quality culture rather than individual quality practices is required in organizations for the successful execution of quality improvement efforts. This research contributes to the body of knowledge by analyzing the relationship between QM in pharmaceutical SCs and SCP, which is previously unexplored. In doing so, it provides a useful perspective from a developing country's point of view that makes this chapter not only useful for managers of pharmaceutical companies in Pakistan but also to those working in the same sector elsewhere in the developing countries. It will help the practitioners in identifying the quality practices that impact the SCP. These results may inspire managers to take decisions about implementation or continuation of the quality programs in their organizations. Over the last decade or so, quality movement has lost its momentum in Pakistan and ISO 9000 remains to be the only manifestation of QM in organizations. Whereas ISO standards tell the organizations whether they are doing the *things right*, they do not specify the *right things* to do. The significant relationship between

QM and SCP indicates that QM throughout the SCs is relevant to modern organizations' success and needs to be considered as a viable option for creating competitive advantage.

Very few studies have sought to explain the impact of QM on performance in the context of Pakistan. Most studies that have done so are sector specific including the current research. More evidence across sectors is required in order to substantiate the results of handful of studies that have proven the relationship between QM and various aspects of organizational performance in Pakistan. Field of SCM is in its infancy stage and very little work has been done on SCP measurement in Pakistan. Studies are required to identify performance matrix for SCs in various industries. In terms of pharmaceutical industry, it would be interesting to find out if the factors affecting the SCP are same or different across cultures. This would require further studies in other cultural contexts in pharmaceutical manufacturing sector.

References

Agarwal, A. and Shankar, R. (2002). Analyzing alternatives for improvement in supply chain performance. *Work Study*, *51*(1), 32–37.

Ahire, S. L., Golhar, D. Y., and Waller, M. A. (1996). Development and validation of TQM implementation constructs. *Decision Sciences*, *27*(1), 23–56.

Ahsan, A. (2010). Barriers against implementing and managing quality in Information Technology Industry of Pakistan. *Nucleus*, *47*(2), 91–123.

Asamoah, M. D., Abor, M. P. A., and Opare, M. M. A. K. (2011). An examination of pharmaceutical supply chain for artemisinin-based combination therapies in Ghana. *Management Research Review*, *34*(7), 790–809.

Awan, H., Bukhari, K., and Qureshi, M. A. (2008). Critical success factors of TQM: Impact on business performance of manufacturing sector in Pakistan. *International Journal of Business and Management*, *1*(2), 187–203.

Awan, M. U., Raouf, A., Ahmad, N., and Sparks, L. (2009). Total quality management in developing countries: A case of pharmaceutical wholesale distribution in Pakistan. *International Journal of Pharmaceutical and Healthcare Marketing*, *3*(4), 363–380.

Beamon, B. M. (1999). Measuring supply chain performance. *International Journal of Operations & Production Management*, *19*(3), 275–292.

Beamon, B. M. and Ware, T. M. (1998). A process quality model for the analysis, improvement and control of supply chain systems. *Logistics Information Management*, *11*(2), 105–113.

Brewer, P. C. and Speh, T. W. (2000). Using the balanced scorecard to measure supply chain performance. *Journal of Business Logistics*, *21*(1), 75–94.
Carr, A. S., Leong, G. K., and Sheu, C. (2000). A study of purchasing practices in Taiwan. *International Journal of Operations & Production Management*, *20*(12), 1427–1446.
Casadesús, M. and de Castro, R. (2005). How improving quality improves supply chain management: Empirical study. *The TQM Magazine*, *17*(4), 345–357.
Chan, F. T. S. and Qi, H. (2003). Feasibility of performance measurement system for supply chain: A process-based approach and measures. *Integrated Manufacturing Systems*, *14*(3), 179–190.
Chandra, C. and Kumar, S. (2000). Supply chain management in theory and practice: A passing fad or a fundamental change? *Industrial Management & Data Systems*, *100*(3), 100–114.
Choi, T. Y. and Rungtusanatham, M. (1999). Comparison of quality management practices: Across the supply chain and industries. *Journal of Supply Chain Management*, *35*(1), 20–27.
Cronbach, L. J. (1951). Coefficient alpha and the internal structure of tests. *Psychometrika*, *16*(3), 297–334.
Das, A., Pagell, M., Behm, M., and Veltri, A. (2008). Toward a theory of the linkages between safety and quality. *Journal of Operations Management*, *26*(4), 521–535.
Das, K. (2011). A quality integrated strategic level global supply chain model. *International Journal of Production Research*, *49*(1), 5–31.
Douglas, T. J. and Judge, W. Q. (2001). Total quality management implementation and competitive advantage: The role of structural control and exploration. *The Academy of Management Journal*, *44*(1), 158–169.
Epsicom. (2011). The Pharmaceutical Market: Pakistan. Retrieved September 2, 2011, from http://www.espicom.com/prodcat2.nsf/Product_ID_Lookup/00000314?OpenDocument.
Field, A. P. (2009). *Discovering Statistics using SPSS: And Sex and Drugs and Rock 'n' Roll* (3rd edn.). SAGE Publications Ltd., London, U.K.
Flynn, B. and Flynn, E. (2005). Synergies between supply chain management and quality management: Emerging implications. *International Journal of Production Research*, *43*(16), 3421–3436.
Flynn, B., Schroeder, R. G., and Sakakibara, S. (1995). The impact of quality management practices on performance and competitive advantage. *Decision Sciences*, *26*(5), 659–691.
Forker, L., Mendez, D., and Hershauer, J. (1997). Total quality management in the supply chain: What is its impact on performance? *International Journal of Production Research*, *35*(6), 1681–1702.
Forslund, H. and Jonsson, P. (2007). The impact of forecast information quality on supply chain performance. *International Journal of Operations & Production Management*, *27*(1), 90–107.
Forza, C. and Filippini, R. (1998). TQM impact on quality conformance and customer satisfaction: A causal model. *International Journal of Production Economics*, *55*(1), 1–20.

Foster Jr., S. T. (2008). Towards an understanding of supply chain quality management. *Journal of Operations Management*, *26*(4), 461–467.

Foster Jr., S. T., Wallin, C., and Ogden, J. (2010). Towards a better understanding of supply chain quality management practices. *International Journal of Production Research*, *49*(1), 1–16.

Fotopoulos, C. V. and Psomas, E. L. (2010). The structural relationships between TQM factors and organizational performance. *The TQM Journal*, *22*(5), 539–552.

Fynes, B., De Burca, S., and Voss, C. (2005). Supply chain relationship quality, the competitive environment and performance. *International Journal of Production Research*, *43*(16), 3303–3320.

Gunasekaran, A., Patel, C., and Tirtiroglu, E. (2001). Performance measures and metrics in a supply chain environment. *International Journal of Operations & Production Management*, *21*(1/2), 71–87.

Hair Jr., J. F., Anderson, R. E., Tatham, R. L., and Black, W. C. (2007). *Multivariate Data Analysis: With Readings* (6th edn.). Prentice-Hall, Inc.

Hausman, W. (2004). Supply chain performance metrics. In *The Practice of Supply Chain Management: Where Theory and Application Converge.* Springer, pp. 61–73.

Kannan, V. and Tan, K. (2007). The impact of operational quality: A supply chain view. *Supply Chain Management: An International Journal*, *12*(1), 14–19.

Kaynak, H. (2003). The relationship between total quality management practices and their effects on firm performance. *Journal of Operations Management*, *21*, 405–435.

Kaynak, H. and Hartley, J. L. (2008). A replication and extension of quality management into the supply chain. *Journal of Operations Management*, *26*(4), 468–489.

Khan, A. (2010). Evaluating the Deming management model of total quality in telecommunication industry in Pakistan—An empirical study. *International Journal of Business and Management*, *5*(9), P46.

Kuei, C. and Madu, C. (2001). Identifying critical success factors for supply chain quality management. *Asia Pacific Management Review*, *6*(4), 409–423.

Kuei, C., Madu, C., and Lin, C. (2001). The relationship between supply chain quality management practices and organizational performance. *International Journal of Quality & Reliability Management*, *18*(8), 864–872.

Kuei, C., Madu, C. N., and Lin, C. (2008). Implementing supply chain quality management. *Total Quality Management & Business Excellence*, *19*(11), 1127–1141.

Kuei, C., Madu, C. N., and Lin, C. (2010). Developing global supply chain quality management systems. *International Journal of Production Research*, *1*, 25.

Kuei, C., Madu, C. N., Lin, C., and Chow, W. S. (2002). Developing supply chain strategies based on the survey of supply chain quality and technology management. *International Journal of Quality & Reliability Management*, *19*(7), 889–901.

Kureshi, N., Qureshi, F., and Sajid, A. (2010). Current health of quality management practices in service sector SME. *The TQM Journal*, *22*(3), 317–329.

Lai, K., Cheng, T., and Yeung, A. C. L. (2005). Relationship stability and supplier commitment to quality. *International Journal of Production Economics*, *96*(3), 397–410.

Lambert, D. M. and Pohlen, T. L. (2001). Supply chain metrics. *The International Journal of Logistics Management*, *12*(1), 1–19.

Lee, C. W., Kwon, I. W. G., and Severance, D. (2007). Relationship between supply chain performance and degree of linkage among supplier, internal integration, and customer. *Supply Chain Management: An International Journal*, *12*(6), 444–452.

Lin, C., Chow, W. S., Madu, C. N., Kuei, C. H., and Pei Yu, P. (2005). A structural equation model of supply chain quality management and organizational performance. *International Journal of Production Economics*, *96*(3), 355–365.

Mohrman, S. A., Tenkasi, R. V., Lawler, E. E., and Ledford, G. E. (1995). Total quality management: Practice and outcomes in the largest US firms. *Employee Relations*, *17*(3), 26–41.

Nair, A. (2006). Meta-analysis of the relationship between quality management practices and firm performance—Implications for quality management theory development. *Journal of Operations Management*, *24*(6), 948–975.

Obaid, A. (2009). Quality of ceftriaxone in Pakistan: Reality and resonance. *Pakistan Journal of Pharmaceutical Sciences*, *22*(2), 220.

Ou, C. S., Liu, F. C., Hung, Y. C., and Yen, D. C. (2010). A structural model of supply chain management on firm performance. *International Journal of Operations & Production Management*, *30*(5), 526–545.

Pakistan Pharmaceutical Manufacturers Association. (2011). Retrieved September 2, 2011, from http://www.ppma.org.pk/PPMAIndustry.aspx.

Powell, T. C. (1995). Total quality management as competitive advantage: A review and empirical study. *Strategic Management Journal*, *16*(1), 15–37.

Prajogo, D. I. and Sohal, A. S. (2003). The relationship between TQM practices, quality performance, and innovation performance: An empirical examination. *International Journal of Quality & Reliability Management*, *20*(8), 901–918.

Prajogo, D. I. and Sohal, A. S. (2006). The relationship between organization strategy, total quality management (TQM), and organization performance—The mediating role of TQM. *European Journal of Operational Research*, *168*(1), 35–50.

Ramos, J., Asan, S., and Majetic, J. (2007). Benefits of applying total quality management techniques to support supply chain management. Paper presented at the *International Logistics and Supply Chain Congress*, November 8–9, 2007.

Raouf, A. and Yusaf, I. (2011). *Quality Management: A Customer Driven Approach*. University of Management and Technology Press, Lahore, Pakistan.

Robinson, C. J. and Malhotra, M. K. (2005). Defining the concept of supply chain quality management and its relevance to academic and industrial practice. *International Journal of Production Economics*, *96*(3), 315–337.

Romano, P. and Vinelli, A. (2001). Quality management in a supply chain perspective: Strategic and operative choices in a textile-apparel network. *International Journal of Operations & Production Management*, *21*(4), 446–460.

Saad, M. and Patel, B. (2006). An investigation of supply chain performance measurement in the Indian automotive sector. *Benchmarking: An International Journal*, *13*(1/2), 36–53.

Samson, D. and Terziovski, M. (1999). The relationship between total quality management practices and operational performance. *Journal of Operations Management*, *17*(4), 393–409.

Sezen, B. (2008). Relative effects of design, integration and information sharing on supply chain performance. *Supply Chain Management: An International Journal*, *13*(3), 233–240.

Shahrukh, N. (2011). Pharmaceutical industry walking a tightrope. *Weekly Pulse*. Retrieved February 18, 2011, from http://www.weeklypulse.org/details.aspx?contentID=231&storylist=12.

Sila, I. and Ebrahimpour, M. (2003). Examination and comparison of the critical factors of total quality management (TQM) across countries. *International Journal of Production Research*, *41*(2), 235–268.

Sila, I. and Ebrahimpour, M. (2005). Critical linkages among TQM factors and business results. *International Journal of Operations & Production Management*, *25*(11), 1123–1155.

Sila, I., Ebrahimpour, M., and Birkholz, C. (2006). Quality in supply chains: An empirical analysis. *Supply Chain Management: An International Journal*, *11*(6), 491–502.

Soltani, E., Azadegan, A., Liao, Y. Y., and Phillips, P. (2011). Quality performance in a global supply chain: Finding out the weak link. *International Journal of Production Research*, *49*(1), 269–293.

Srivastava, S. (2008). Towards estimating Cost of Quality in supply chains. *Total Quality Management & Business Excellence*, *19*(3), 193.

Tabachnick, B. G., Fidell, L. S., and Osterlind, S. J. (2007). *Using Multivariate Statistics* (5th edn.). Allyn & Bacon, Boston, MA.

Tan, K. C., Kannan, V. R., Handfield, R. B., and Ghosh, S. (1999). Supply chain management: An empirical study of its impact on performance. *International Journal of Operations & Production Management*, *19*(10), 1034–1052.

Terziovski, M. and Samson, D. (1999). The link between total quality management practice and organisational performance. *International Journal of Quality & Reliability Management*, *16*(3), 226–237.

Van Hoek, R. I. (1998). "Measuring the unmeasurable"—Measuring and improving performance in the supply chain. *Supply Chain Management: An International Journal*, *3*(4), 187–192.

Vanichchinchai, A. and Igel, B. (2009). Total quality management and supply chain management: Similarities and differences. *The TQM Journal*, *21*(3), 249–260.

Vanichchinchai, A. and Igel, B. (2010). The impact of total quality management on supply chain management and firm's supply performance. *International Journal of Production Research*, *49*(1), 1–20.

Yeung, A. C. L. (2008). Strategic supply management, quality initiatives, and organizational performance. *Journal of Operations Management*, *26*(4), 490–502.

Zaheer, F. (2011). Research in pharmaceutical sector: Pakistani companies lag behind their Indian, Chinese counterparts *Tribune*. Retrieved August 15, 2011, from http://tribune.com.pk/story/231549/research-in-pharmaceutical-sector-pakistani-companies-lag-behind-their-indian-chinese-counterparts/.

Zelbst, P. J., Green Jr., K. W., Sower, V. E., and Reyes, P. (2009). Impact of supply chain linkages on supply chain performance. *Industrial Management & Data Systems*, *109*(5), 665–682.

Zhang, L., Wang, S., Li, F., Wang, H., Wang, L., and Tan, W. (2011). A few measures for ensuring supply chain quality. *International Journal of Production Research*, *49*(1), 87–97.

2

Risk Analysis and Efficient Logistics for Maritime Ports and Waterways in Qatar

SELIM BORA, GINO J. LIM,
AND HAMID R. PARSAEI

Contents

2.1 Introduction

Due to the undesired implications of maritime mishaps such as ship collisions and their consequent damages to maritime assets, the safety and security of waterways, ports, and other maritime assets are of utmost importance to authorities and researches. Terrorist attacks, piracy, accidents, and environmental damages are some of the concerns. With the expansion of ports and port operations within Qatar such as Ras Laffan and Mesaieed, the construction of new ones such as the new port in Doha, Qatar, needs to consider the risks its ports are facing and address them. The overall goal of this chapter is to provide guidelines for maritime risk assessment for ports in Qatar. This study provides a detailed literature review on different threats and their consequences pertinent to the maritime industry and discusses various risk assessment models. The methods are all explained and a framework for maritime risk analysis studies for ports of Qatar is established.

2.2 Background: Economics and Ports of Qatar

Qatar Statistics Authority reports a compound annual growth rate of 27.5% from 2004 to 2011 in nominal GDP. Liquefied natural gas (LNG) is the main driving force behind this growth. Main exports of Qatar are LNG (60% of total exports) and crude oil (30%), being delivered to Japan (28% of total exports), South Korea (19%), and India (11%) mainly. Exports in Qatar were reported to be 123,971 QAR million in the third quarter of 2013 by the Qatar Central Bank as can be seen in Figure 2.1.

Imports in Qatar on the other hand were reported to be 28,201 QAR million in the third quarter of 2013 by the Qatar Central Bank as can be seen in Figure 2.2 with the following breakdown: transport equipment and parts (19% of total imports); nuclear reactors, boilers, machinery, and mechanical appliances and parts (15%); base metals and articles thereof (11%); and electrical machinery and equipment and parts (10.7%), being delivered from the United States (11% of total imports), China (10%), Japan (8.2%), United Arab Emirates (8%), and Germany (7%) mainly.

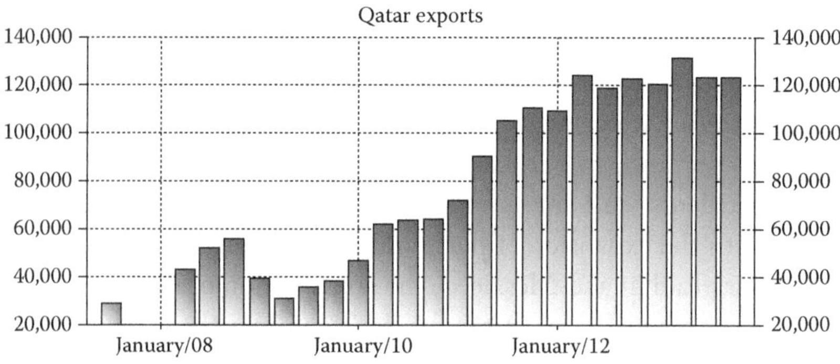

Figure 2.1 Qatar exports by month. (From Trading Economics, Qatar exports, 2013a, http://www.tradingeconomics.com/qatar/exports, accessed December 10, 2013.)

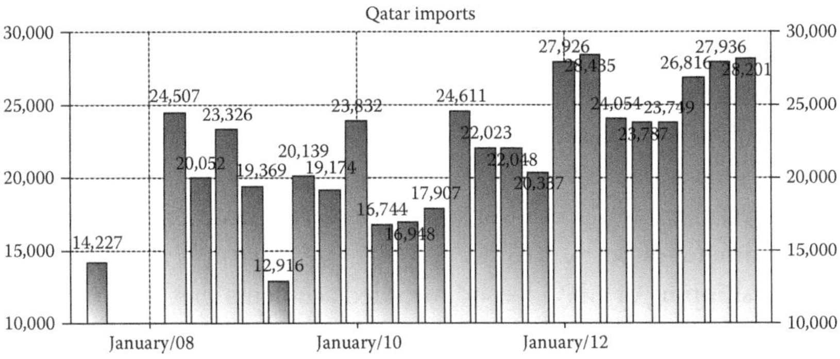

Figure 2.2 Qatar imports by month. (From Trading Economics, Qatar imports, 2013b, http://www.tradingeconomics.com/qatar/imports, accessed December 10, 2013.)

As a result of the growth in the volume of imports and exports, ports of Qatar have experienced an increased trend in traffic in recent years. Currently, there are three main ports in Qatar: Ras Laffan, Doha, and Mesaieed. Ras Laffan and Mesaieed are mainly dedicated to LNG and crude oil traffic, whereas Doha is a container terminal.

The numbers of arriving and departing vessels are increasing every year, and a reported gross tonnage of vessels docked increased by 14.8% from 2010 to 2011. The increase in net tonnage was reported to be 16.8% from 2010 to 2011. The main types of vessels were oil and gas tankers amounting to 2694 vessels with a tonnage of 230.1 million in 2011, an increase of 17.4% compared with 2010 figures. Table 2.1 shows the vessel movements from and to the main ports

Table 2.1 Vessel Traffic in Main Ports of Qatar

	2010			2011		
PORT	NUMBER OF VESSELS	GROSS TONNAGE (MN. TONS)	NET TONNAGE (MN. TONS)	NUMBER OF VESSELS	GROSS TONNAGE (MN. TONS)	NET TONNAGE (MN. TONS)
Doha	1144	19.0	7.1	1188	19.8	7.4
Mesaieed	2161	44.7	23.4	2112	46.2	24.1
Ras Laffan	1719	153.8	58.9	1944	190.9	77.5
Total	5159	237.6	101.9	5353	272.7	119.0

Source: Ministry of Development Planning and Statistics, Qatar, With gross tonnage of 272.7 million, 5353 vessels docked at Qatari Ports in 2011, 2013, http://www.qsa.gov.qa/eng/News/2013/Articles/13.htm, accessed December 10, 2013.

Note: Mn., million.

of Qatar. It should be noted that the vessel traffic in Ras Laffan has increased by a higher percentage than the other two ports. With the additional capacity it has, the traffic will continue to grow. Currently, Ras Laffan has six active LNG berths and three additional berths that will become active over the upcoming years. As the port traffic increases, there is a greater chance of vessel-related accidents within the perimeter of ports.

The shipping of LNG has no recorded accidents that have resulted in loss of cargo. With the strong infrastructure of LNG ships, even in the case of an accident, loss of cargo is unlikely. However, collisions or terrorist attacks may still lead to leaks or fires, or both. Pitbaldo et al. (2005) report two cases in which release of LNG is possible, one being as a result of a 90° angle collision with another vessel and the other one being as a result of a terrorist attack. After a 90° angle collision, the hazard distance is the worst-case scenario, while the LNG may not ignite immediately. This may allow a vapor cloud to form and disperse downwind to an extent of 920 m. The vapor can ignite and flash back to the source of LNG spill, where a pool fire with a range of 440 m would form. In the case of terrorist attack, immediate ignition is most likely, and the pool fire range goes up to 750 m.

Even though these scenarios are less likely to occur at LNG terminals of ports in Qatar due to their robust infrastructure and secure boundaries, LNG spillage is an undesirable event. With the increasing port traffic at Qatar, there is a clear need to understand and develop preventive measures to avoid LNG-related disasters. Maritime risk analysis studies are a good way to tackle this problem, which are discussed in the following sections. Section 2.3 talks about existing work on maritime risk analysis and summarizes the guidelines for a successful risk analysis. Section 2.4 reviews the literature on maritime simulation and provides a basic input for a simulation study on risk analysis for ports of Qatar.

2.3 Risk Analysis

Risk management process involves several steps. These are (1) identifying hazards; (2) assessing the risks, that is, assess the likelihood and consequences, as well as classifying the risks; (3) controlling the risks; and (4) monitoring the control measures. A hazard is something that has the potential to cause harm to people, property, or the environment. A risk is the probability of a hazard causing harm to people, property, or the environment, and control is a mechanism designed to minimize the risk of a hazard happening. These concepts will be talked in more detail during the remainder of this section as well as the rest of this chapter, focusing more on maritime risk analysis.

The importance of maritime transportation to the world economy cannot be overemphasized. It is widely accepted that more than 90% of the world's international trade travels by sea. As such, global economic interdependency among nations is largely reliant on the success of the maritime industry. Unlike other modes of transportation, maritime transportation has proved to be the most cost-effective way of transporting bulk goods, petroleum products, food supplies, manufactured goods, containerized cargo, etc., over long distances. On the other hand, maritime transportation could be described as the most dangerous means of transportation. Piracy, inclement weather conditions, natural disasters (tsunamis, earthquakes, etc.), narrow waterways, dangerous uncharted waterways, and vessel collisions are some of the identified threats to the safety and security of vessels, commodity, passengers, and seafarers. International maritime regulations adopted by the industry have, to a great extent, improved safety and security in the maritime industry. However, better decision making can be achieved if risk exposures can be accurately determined (and adequate measures proffered to mitigate the effects).

The shipping safety regime consists primarily of international safety codes and regulations issued by the International Maritime Organization (IMO) and rules for the construction of ships issued by independent classification societies. Marine safety regulations have grown in a mainly reactive way, with accident experience providing the prime motivation for improved regulation. This approach was successful for large fleets of similar ships, in which past experience formed a good basis for safety management. However, it has been less effective for unusual and rapidly changing designs, such as many offshore installations and several important types of ships. In response, the shipping industry is developing formal safety assessment as a more proactive approach to regulation. The advantage of marine regulations is that they encapsulate the accumulated wisdom from accident experience and from the judgment of many experts worldwide who have contributed to refining and improving them. The disadvantage when performing a risk assessment is that the accident experience and anticipated hazards that underpinned each rule are not recorded, and so it is very difficult to tell how safety critical a particular rule might be for a particular installation.

Based on this background, it is only natural that extensive research efforts should be focused on the safety and security of maritime transportation assets: vessels, ports, and waterways. In maritime port systems, accident data related to port operations are often nonexistent. As such, estimation of accident probabilities in ports usually necessitates analysis of the opinions of individuals with domain knowledge of important maritime operations along with mathematical models. Also, maritime port situations in a port are constantly evolving due to different traffic rules, changing traffic patterns, and other conditions such as visibility and wind. Based on the foregoing, simulation, mathematical modeling, and expert judgment elicitation play very important roles in modeling maritime security and safety risks in ports and waterways.

Research papers with quantitative views use various solution methods, which can be classified into three primary categories. Because of the relative lack of data in the field, a noticeable number of these papers adopt the simulation approach using commercial simulation packages such as Arena or building their own simulation programs. Statistical methods have also been used in a broad spectrum, including probability models, Bayesian belief network modeling, regression, queuing theory, kernel density technique, FSA, and HFACS tool, to name a few. Furthermore, fuzzy approach was developed to deal with quantitative nature of the expert judgment (Celik et al. 2010). Fault tree analysis approach (Mokhtari et al. 2011) and Monte Carlo techniques (Jansson and Gustafsson 2008) can be other useful techniques for risk analysis. Finally, a very limited number of the reviewed papers take either linear or mixed integer programming approach in order to optimize their objective functions.

Iakovou (2001) presents a multiobjective network flow concerning environmental pollution resulted from oil spills, and Ghafoori and Altiok (2012) propose an optimization model for sensor placement. Combination of the aforementioned methods has been also reported in the literature.

Other notable approaches have also been reported that include more qualitative and semiqualitative techniques. Preliminary hazard analysis (PHA) is a semiquantitative analysis used to identify all potential hazardous events that may lead to an accident, where these events are ranked according to their severity and then potential preemptive and follow-up actions are identified. It is more intended for use as an initial risk study at earlier stages of a project or complete risk analysis of a rather simple system. Failure mode effects analysis (FMEA) is single point of failure analysis, where for each component, failure modes and their resulting effects on the entire system are recorded. The hazard and operability (HAZOP) study consists of systematically analyzing all the deviations of the operating parameters of the different elements or of the stages of the operating mode and in analyzing those that can potentially lead to a dangerous event (Flaus 2013). Event tree analysis (ETA) explores responses from a single initiating event for assessing probabilities of possible outcomes. Bow-tie diagram analysis is similar to ETA, where diagrammatic representations of hazardous events are used for risk analysis, which is usually required by governmental regulators. Human reliability analysis (HRA) is a tool used for assessing the resilience of systems based on human factors and ergonomics. Layer of protection analysis (LOPA) is a tool for assessing the effectiveness of protection measures used to mitigate risk. It provides a semiquantitative evaluation of the frequency of potential incidents and the probability of the protection measures failing.

As stated before, all these approaches are used for risk assessment as an aid to the decision making, addressing financial risks, health risks, safety risks, environmental risks, and other types of business risks associated with a decision. An appropriate analysis of these risks will provide information that is critical to good decision making and often will clarify the decision to be made. The information generated through risk assessment can often be communicated to the organization to help impacted parties understand the factors that influenced the decision.

Risk assessment is performed in several systematic ways: (1) hazard identification, (2) frequency assessment, (3) consequence assessment, and (4) risk evaluation. The level of information needed to make a decision varies widely. In some cases, after identifying the hazards, qualitative methods of assessing frequency and consequence are satisfactory to enable the risk evaluation. In other cases, a more detailed quantitative analysis is required. There are many different analysis techniques and models that have been developed to aid in conducting risk assessments. A key to any successful risk analysis is choosing the right method (or combination of methods) for the situation at hand. This study reviews common methods used in risk analysis for maritime traffic.

The risk of an accident is the probability of occurrence of the accident and the consequences associated. An incident triggers an unsafe condition that may result in an accident. Typically, incidents are caused by instigators (human error, mechanical error, etc.) in situations (lack of visibility, water and traffic conditions, etc.) prone to accidents or man-made incidents. There are a variety of deleterious events, such as accidents (collisions, groundings, spills, and others), their instigators (human error, steering failure, navigational failure, and others), their consequences (human casualty, environmental damage, and others), and situational factors (current, tide, visibility, and others); thus, a causal chain relationship needs to be set up as a preamble before a risk model is developed. The combination of a triggering event and situational conditions (location, wind, and weather) yields a potential for an unwanted event.

Typically, incidents are caused by instigators (human error, mechanical error, etc.) in situations (lack of visibility, water and traffic conditions, etc.) prone to accidents or man-made incidents. There are a variety of deleterious events, such as accidents (collisions, groundings, spills, and others), their instigators (human error, steering failure, navigational failure, and others), their consequences (human casualty, environmental damage, and others), and situational factors (current, tide, visibility, and others); thus, a causal chain relationship needs to be set up as a preamble before a risk model is developed as in Figure 2.3.

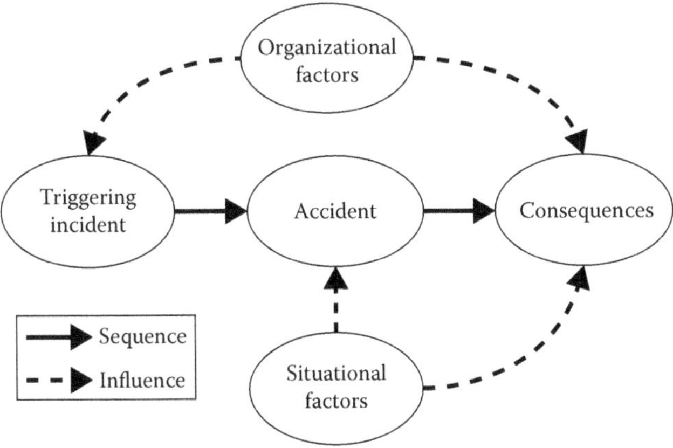

Figure 2.3 Illustration of conditional relationships. (From Harrald, J.R. et al., *Safety Sci.*, 30(1), 235, 1998.)

Table 2.2 Attributes Affecting Risk Previously Studied in the Literature

ORGANIZATIONAL FACTORS	SITUATIONAL FACTORS
Vessel size	Location
Vessel age	Vessel proximity
Vessel type	Population in the proximity
Vessel flag	Traffic direction
Vessel cargo	Traffic density
Pilot request	Escort vessels
Officer experience	Wind speed/direction
	Infrastructure
	Visibility
	Current
	Time of the day

It is essential to study the organizational and situational attributes affecting the risks. Table 2.2 summarizes these attributes that have previously been addressed in the literature.

Next, the causal relationships between instigators, incidents, and consequences must be studied as in Tables 2.3 and 2.4. In order to determine casual relationship between instigators and incidents, it should be considered that not all the instigators can be a reason for an incident, and not all the incidents can be a result of an instigator. The same idea is true for the casual relationship between incidents and consequences.

Table 2.3 Casual Relationship between Instigators and Incidents

	COLLISION	OIL SPILL	TERRORIST ATTACK	FIRE/ EXPLOSION	GROUNDING	RAMMING	SINKING
Human error	✓	✓		✓	✓	✓	✓
Terrorist act	✓		✓	✓			✓
Steering failure	✓	✓			✓	✓	✓
Propulsion failure	✓	✓			✓	✓	✓
Communication failure	✓				✓	✓	✓
Mechanical failure		✓		✓			✓

Table 2.4 Casual Relationship between Incidents and Consequences

	HUMAN CASUALTY	ENVIRONMENTAL DAMAGE	PROPERTY DAMAGE	TRAFFIC EFFECTIVENESS	POLITICAL ISSUES
Collision	✓	✓	✓	✓	
Oil spill	✓	✓		✓	✓
Terrorist attack	✓	✓	✓	✓	✓
Fire/explosion	✓	✓	✓	✓	
Grounding		✓		✓	
Ramming	✓	✓	✓	✓	
Sinking	✓	✓	✓	✓	

Risk modeling, which is a process, should be handled under generally accepted guidelines for risk assessment for credibility and success. The process should include (1) risk identification, (2) risk quantification and measurement, (3) risk evaluation, and (4) risk mitigation (Grabowski et al. 2000).

A risk framework such as that provided in Figures 2.4 and 2.5 is a must in risk modeling and is usually the first step. A risk framework should provide (1) definition of risks; (2) definitions and examples for components of the error chain; (3) descriptions of accidents, incidents, and unusual events; and (4) identification of risk mitigation measures categorized by their impact on the error chain (Grabowski et al. 2000). Once the framework is established, the following tasks are required for a credible risk analysis.

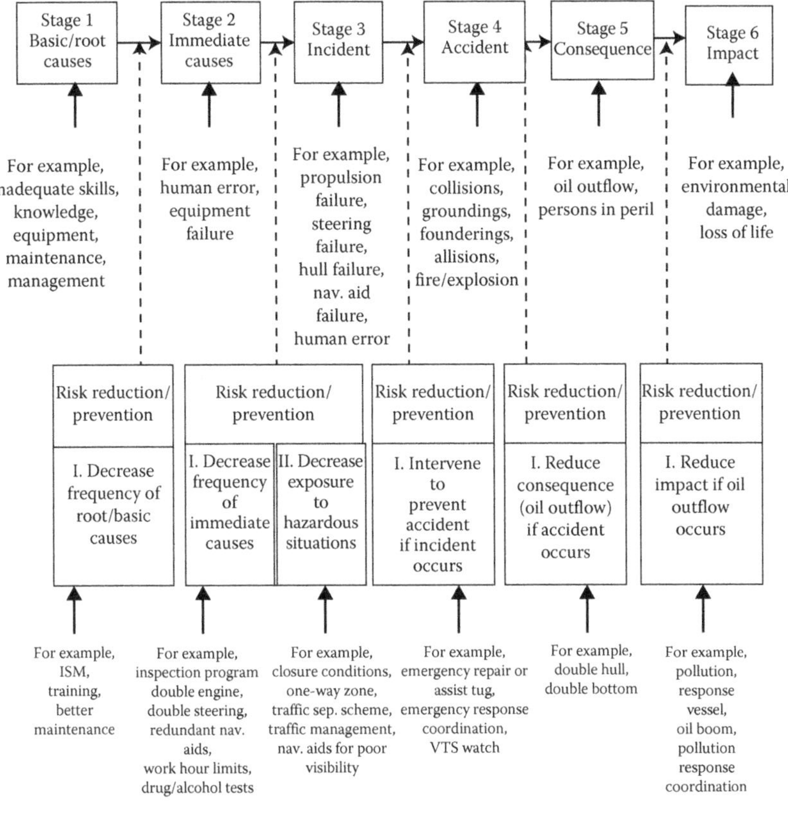

Figure 2.4 Framework for maritime risk assessment and risk reduction interventions. (From Harrald, J.R. et al., *Safety Sci.*, 30(1), 235, 1998.)

2.3.1 Historical System Benchmarks

One of the first tasks is to develop benchmarks on risk levels of the system to analyze historic and present performance. This also helps in identifying patterns of incident and accident occurrence in the system, which is useful for proposing risk mitigation measures. Past data analysis also can help in identification of latent pathogens, incubation periods, and catalysts in the system. The use of accident data for comparing performance can be problematic. The number of accidents may be low or there could be missing data, making it hard to analyze the interactions between factors.

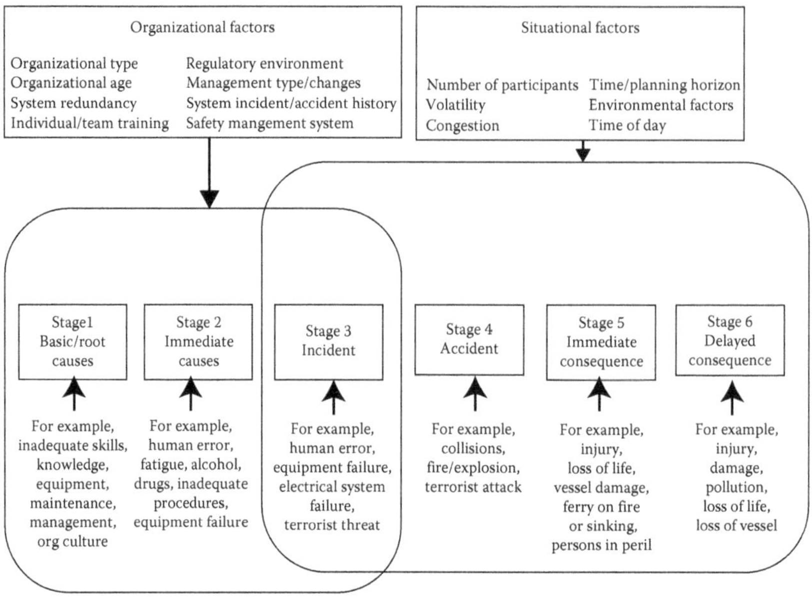

Figure 2.5 Risk event error chain. (From Grabowski, M. et al., *IEEE Trans. Syst. Man Cybernet., Part A: Syst. Humans*, 30(6), 651, 2000.)

2.3.2 Dynamic Risk Models

The dynamic risk modeling allows capturing of the dynamic nature of risk and assessing risk migration. It includes steps to identify the series of events leading to accidents, estimation of the probabilities of these events, and evaluation of the consequences of different degrees of system failure. Finally, the impacts of risk mitigation measures on levels of risk in the system can be estimated.

2.3.3 Assessment of the Role of Human and Organizational Error

Formal assessments of human and organizational error can capture important performance parameters, and risk mitigation measures to reduce risk can be proposed accordingly. The errors can be identified and then categorized using the human and organizational error taxonomy developed by Reason (1997) and illustrated in Figure 2.6. Estimates of the conditional probabilities that link the stages in the causal chain must be made in order to predict the risk of accidents due to human and organizational error.

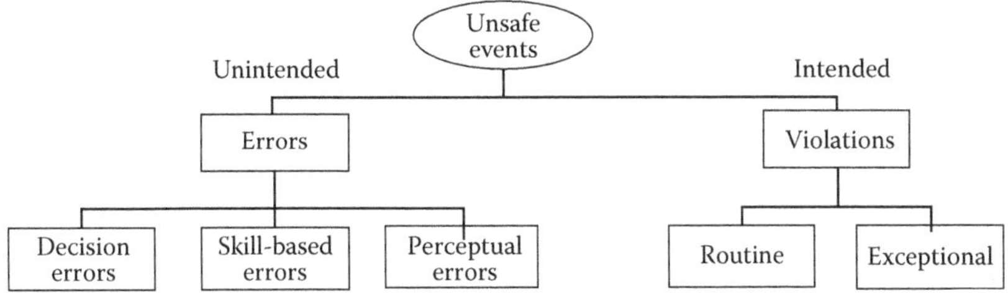

Figure 2.6 Human and organizational errors. (From Reason, J.T., *Managing the Risks of Organizational Accidents*, Vol. 6, Ashgate, Aldershot, U.K., 1997.)

2.3.4 Domain-Specific Models

They can be used to address any special risk requirements in distributed, large-scale systems, that is, consequences of a collision. Goerlandt et al. (2012) is on collisions and Gucma and Przywarty (2007) is on oil spills. Ors (2003) simulates the spill of oil, and this operation is carried out by the Lagrangian discrete particle tracking method using output from the hydrodynamic model for the velocity field. Projections of an oil spill occurrence are presented on an hourly basis.

As will be discussed in Section 2.4, simulation is a powerful tool that can be used for many purposes for risk analysis, such as establishing historical system benchmarks, dynamic risk models, and domain-specific models. Simulation modeling has been used in various fields where analytical models cannot be used due to the complex nature of problems. Even though simulation is used for modeling operations related to terminal operations as well, Section 2.4 focuses on the use of simulation for maritime studies.

2.4 On Maritime Simulation

Simulation studies in maritime transportation domain can be categorized under applications on port/terminal operations and logistics, modeling of vessel traffic on waterways for scenario and policy analysis, and using simulation platforms as a tool to evaluate accident probabilities, risks, and various economic and technical issues. In this section, use of simulation for maritime risk analysis will be discussed.

In a marine environment, traffic patterns change over time in a complex manner. Simulation can provide critical input to decision makers in analyzing future scenario variations in maritime transportation systems, as well as evaluating risks and consequences of countermeasures.

Case studies are built to accurately represent the operation within the study domain. Vessels, environmental conditions, and traffic regulations are built into the model for a realistic simulation. Using the simulation model, a counting model is developed to observe and record snapshots of the study area at regular intervals and count the occurrences of interest.

Literature on simulation modeling of vessel traffic on waterways is not large but growing. Golkar et al. (1998) develop a simulation model for the Panama Canal as a tool for scenario and policy analyses. Thiers and Janssens (1998) develop a detailed maritime traffic simulation model for the port of Antwerp, Belgium, including navigation rules, tides, and lock operations in order to investigate the effects of a container quay to be built outside the port on the vessel traffic and especially on the waiting time of the vessels. Merrick et al. (2003) perform traffic density analysis, which would lead later to the risk analysis for the ferry service expansion in the San Francisco Bay area. They try to estimate the frequency of vessel interactions using a simulation model, in which vessel movements, visibility conditions, and geographical features are included. Cortés et al. (2007) simulate both the freight traffic and terminal logistics for the port of Seville, Spain, using Arena software focusing on port utilization. Smith et al. (2007) work on congestion in Upper Mississippi River through building a traffic simulation model and test different operating conditions. For the strait of Istanbul, there is considerable literature bringing different perspectives in which simulation modeling was used for scenario and policy analyses. Köse et al. (2003) work with an elementary model of the strait of Istanbul and test the effect of arrival intensity on waiting times. Özbaş and Or (2007) and Almaz et al. (2006) develop extensive simulation models including vessel types, cargo characteristics, pilot and tugboat services, traffic rules, and environmental conditions and investigated effects of numerous factors on different performance measures such as transit times, waiting times, vessel density in the strait, and service utilizations.

In addition to these, in various studies, vessel traffic simulation was inherently used as an environment for further analysis of accident probabilities, risks, and various economic and technical issues. Ince and Topuz (2004) use traffic simulation environment as a test bed for the development of navigational rules and to estimate potential system improvements in the strait of Istanbul. Van Dorp et al. (2001) develop traffic simulations including traffic rules, weather, and relevant environmental conditions for Washington State Ferries in Puget Sound area. Merrick et al. (2002) work on a similar study for the Prince William Sound in order to perform risk assessment through integrating accident probability models. In similar studies, Ulusçu et al. (2009) use a traffic simulator to test and deploy a scheduling algorithm for transit vessels in the strait of Istanbul, and Ulusçu and Altıok (2009) develop a dynamic risk analysis map based on an extensive vessel traffic simulation for the strait of Istanbul. Goerlandt and Kujala (2011) also use vessel traffic simulation to evaluate ship collision probability in the open sea where environmental conditions are negligible. Also, Kujala et al. (2009) study the Gulf of Finland to reduce accident statistics via theoretical modeling. Somanathan et al. (2009) investigate economic viability of Northwest Passage compared to Panama Canal using simulation for vessel movements and environmental conditions. Martagan et al. (2009) build a simulation model to evaluate the performance of rerouting strategies of vessels in the U.S. ports under crisis conditions. On the other hand, Quy et al. (2008) use traffic simulation, which includes tide and wave conditions in order to find optimal channel depths for vessel navigation by minimizing the grounding risk based on a wave-induced ship motion model. Recently, Almaz and Altiok (2012) dealt with simulation modeling of the vessel traffic in Delaware River to study the impact of deepening on performance measures related to navigation and port operations in Delaware River.

2.5 Simulation for Risk Analysis

The simulation is only part of the risk assessment. The incident and accident probabilities require the use of data analysis and expert judgment techniques, while if necessary damage and response time models may have to be created as well (Merrick et al. 2003).

The frequencies of accidents can be predicted using simulation models. Data from the automatic identification system (AIS) can provide an excellent basis for input in creating the simulation model. The types of encounter in maritime are very dependent on the location even though there may be some similarities. Within simulation models, snapshots of the system within certain time intervals are taken to record to the database the number of encounters generated. Similarly, the waterway could be divided into slices, and for each passage of each vessel, a snapshot of the slice could be taken (Ulusçu et al. 2009). Exposure to collision risk is based on the number and type of interactions with other vessels; exposure to grounding risk is based on the time spent in areas where grounding is possible; allision risk exposure is determined by the number of dockings made, and based on the study needs, other methods could be developed.

The system risk at any given time is the summation of the risk posed by each of the vessels in the system. As vessels pass through the system, the waterway and organizational characteristics of the vessels in the system change with time, thus changing the level of risk in the system.

These simulation models can be used to determine the probability of collision or risk of interest for the encounter scenarios, usually based on so-called causation factors, obtained from expert elicitation (Van Dorp et al. 2001), accident statistics (Pedersen 1995), fault tree analysis (Martins and Maturana 2010), Bayesian networks (Hanninen et al. 2012), or near misses (Goerlandt et al. 2012; Montewka et al. 2011). Montewka et al. (2012) come up with a new definition of a ship–ship collision criterion, minimum distance to collision (MDTC), which is related to the shortest distance between two ships at which they must take evasive actions to avoid a collision. Their study allows better probabilistic estimation of accidents and a way to quantify the proportion of the cases in which two ships will collide. Montewka et al. (2010) previously had a similar study dealing with probability of two vessels colliding working with a geometrical model. The determination of the number of collisions occurring in a given time in a given location is done by multiplying the number of collision candidates with the appropriate causation probability, giving the number of collisions.

However, the encounter scenarios do not give reasonable estimates for dynamic parameters, such as collision angle or speed at the moment of impact. This is because of the fact that even if there is a collision, it is rarely observed that there are no evasive maneuvering maneuvers taken prior to collision. Taking no actions prior to accident is referred to as blind navigator.

Another problem with simulation is that random processes control the arrival of vessels and environmental conditions into the simulation. The Bayesian paradigm is widely accepted as a method for dealing with uncertainty. Bayesian modeling can allow for the distinction and handle the underlying differences inherently when used for analyzing data and expert judgments. There is an extensive literature on the theory of Bayesian simulation analysis. Chick (2004) gives an excellent review of Bayesian methods, discussing both input and output uncertainties. They also discuss model uncertainty. However, due to the computational complexity of a Bayesian simulation of a port system with many different types of vessels on many different routes, handling model uncertainty is usually computationally overwhelming.

While some simulation models only estimate these probabilities based on blind navigator assumption, some models go further into detail and build domain-specific models to evaluate encounters leading to collision as in Goerlandt et al. (2012). Once the encounter scenarios are detected or generated, impact scenario models could be used for better understanding of consequences of an accident. Monte Carlo simulation or ship handling simulators are usually used for these purposes.

Using simulation, risk assessment can be made, potential countermeasures can be decided, and their impact can be estimated without interrupting the system. For a risk analysis study to be carried out for ports in Qatar, Table 2.5 is prepared as a starting point for situational attributes that need to be considered based on data on ports of Qatar for basis of simulation.

Table 2.5 Situational Attributes for Ports of Qatar

VARIABLE	SA NAME	NO. OF POSSIBLE SETTINGS	DESCRIPTION
X_1	First vessel type	7	Passenger/cargo/tanker/high speed/tug/yacht/fishing
X_2	Second vessel type	7	Passenger/cargo/tanker/high speed/tug/yacht/fishing
X_3	Tugboat request	2	Yes/no
X_4	Pilot request	2	Yes/no
X_5	Nearest transit vessel proximity (from a given transit vessel)		
X_6	Age	3	New/middle aged/old
X_7	Length (m)	6	0–25 25–50 50–100 100–200 200–300 >300
X_8	Speed	3	Slow/medium/fast
X_9	Flag	47	Countries listed
X_{10}	Flag category	3	Low risk/medium risk/high risk
X_{11}	Visibility	3	<0.5/0.5–1/>1 mile
X_{12}	Current	8	Same direction/opposite direction: 0–2/2–4/4–6/>6 knots/h
X_{13}	Local traffic density		
X_{14}	Vessel reliability	$3 \times 3 = 9$	Age × flag
X_{15}	Time of the day	2	Day/night
X_{16}	Population of the shore		
X_{17}	Property buildings, infrastructures		
X_{18}	Port size	3	Very small/small/medium
X_{19}	Channel depth	3	12.5/14–15.5/24.4 m
X_{20}	Anchorage depth	3	12.5/21.6/23–24.5

References

Almaz, A.O., I. Or, and B. Ozbas. Investigation of the transit maritime traffic in the Strait of Istanbul through simulation modeling and scenario analysis. Department of Industrial Engineering, Boğaziçi University, Istanbul, Turkey, 2006.

Almaz, O.A. and T. Altiok. Simulation modeling of the vessel traffic in Delaware River: Impact of deepening on port performance. *Simulation Modelling Practice and Theory* 22 (2012): 146–165.

Celik, M., S.M. Lavasani, and J. Wang. A risk-based modelling approach to enhance shipping accident investigation. *Safety Science* 48, 1 (2010): 18–27.

Chick, S.E. Bayesian methods for discrete event simulation. In *Proceedings of the 2004 Winter Simulation Conference*, Washington, DC, Vol. 1. IEEE, 2004, pp. 89–100.

Cortés, P., J. Muñuzuri, J.N. Ibáñez, and J. Guadix. Simulation of freight traffic in the Seville inland port. *Simulation Modelling Practice and Theory* 15, 3 (2007): 256–271.

Flaus, J.-M. *Risk Analysis: Socio-Technical and Industrial Systems*. John Wiley & Sons, 2013.

Ghafoori, A. and T. Altiok. A mixed integer programming framework for sonar placement to mitigate maritime security risk. *Journal of Transportation Security* 5, 4 (2012): 253–276.

Goerlandt, F. and P. Kujala. Traffic simulation based ship collision probability modeling. *Reliability Engineering & System Safety* 96 (2011): 91–107.

Goerlandt, F., K. Ståhlberg, and P. Kujala. Influence of impact scenario models on collision risk analysis. *Ocean Engineering* 47 (2012): 74–87.

Golkar, J., A. Shekhar, and S. Buddhavarapu. Panama canal simulation model. In *Proceedings of the 1998 Winter Simulation Conference*, Washington DC, Vol. 2. IEEE, 1998, pp. 1229–1237.

Grabowski, M., J.R.W. Merrick, J.R. Harrold, T.A. Massuchi, and J.D. van Dorp. Risk modeling in distributed, large-scale systems. *IEEE Transactions on Systems, Man and Cybernetics, Part A: Systems and Humans* 30, 6 (2000): 651–660.

Gucma, L. and M. Przywarty. The model of oil spills due to ships collisions in Southern Baltic area. *TransNav, the International Journal on Marine Navigation and Safety of Sea Transportation* 2 (2008): 415–419.

Hänninen, M. and P. Kujala. Influences of variables on ship collision probability in a Bayesian belief network model. *Reliability Engineering & System Safety* 102 (2012): 27–40.

Harrald, J.R., T.A. Mazzuchi, J. Spahn, R. Van Dorp, J. Merrick, S. Shrestha, and M. Grabowski. Using system simulation to model the impact of human error in a maritime system. *Safety Science* 30, 1 (1998): 235–247.

Iakovou, E.T. An interactive multiobjective model for the strategic maritime transportation of petroleum products: Risk analysis and routing. *Safety Science* 39, 1 (2001): 19–29.

Ince, A.N. and E. Topuz. Modelling and simulation for safe and efficient navigation in narrow waterways. *Journal of Navigation* 57, 1 (2004): 53–71.

Jansson, J. and F. Gustafsson. A framework and automotive application of collision avoidance decision making. *Automatica* 44, 9 (2008): 2347–2351.

Köse, E., E. Başar, E. Demirci, A. Güneroğlu, and Ş. Erkebay. Simulation of marine traffic in Istanbul Strait. *Simulation Modelling Practice and Theory* 11, 7 (2003): 597–608.

Kujala, P., M. Hänninen, T. Arola, and J. Ylitalo. Analysis of the marine traffic safety in the Gulf of Finland. *Reliability Engineering & System Safety* 94, 8 (2009): 1349–1357.

Martagan, T.G., B. Eksioglu, S.D. Eksioglu, and A.G. Greenwood. A simulation model of port operations during crisis conditions. In *Proceedings of the 41st conference on Winter simulation*, Austin, TX, 2009, pp. 2832–2843.

Martins, M.R. and M.C. Maturana. Human error contribution in collision and grounding of oil tankers. *Risk Analysis* 30, 4 (2010): 674–698.

Merrick, J.R.W., J.R. Van Dorp, J.P. Blackford, G.L. Shaw, J. Harrald, and T.A. Mazzuchi. A traffic density analysis of proposed ferry service expansion in San Francisco Bay using a maritime simulation model. *Reliability Engineering & System Safety* 81, 2 (2003): 119–132.

Merrick, J.R.W., J.R. Van Dorp, T. Mazzuchi, J.R. Harrald, J.E. Spahn, and M. Grabowski. The Prince William sound risk assessment. *Interfaces* 32, 6 (2002): 25–40.

Ministry of Development Planning and Statistics, Qatar. With gross tonnage of 272.7 million, 5353 vessels docked at Qatari Ports in 2011, 2013. http://www.qsa.gov.qa/eng/News/2013/Articles/13.htm (accessed December 10, 2013).

Mokhtari, K., J. Ren, C. Roberts, and J. Wang. Application of a generic bow-tie based risk analysis framework on risk management of sea ports and offshore terminals. *Journal of Hazardous Materials* 192, 2 (2011): 465–475.

Montewka, J., F. Goerlandt, and P. Kujala. Determination of collision criteria and causation factors appropriate to a model for estimating the probability of maritime accidents. *Ocean Engineering* 40 (2012): 50–61.

Montewka, J., T. Hinz, P. Kujala, and J. Matusiak. Probability modelling of vessel collisions. *Reliability Engineering & System Safety* 95, 5 (2010): 573–589.

Montewka, J., P. Krata, F. Goerlandt, A. Mazaheri, and P. Kujala. Marine traffic risk modelling—An innovative approach and a case study. *Proceedings of the Institution of Mechanical Engineers, Part O: Journal of Risk and Reliability* 225, 3 (2011): 307–322.

Ors, H. Oil transport in the Turkish Straits system: A simulation of contamination in the Istanbul Strait. *Energy Sources* 25, 11 (2003): 1043–1052.

Özbaş, B. and I. Or. Analysis and control of maritime transit traffic through the İstanbul Channel: A simulation approach. *Central European Journal of Operations Research* 15, 3 (2007): 235–252.

Pedersen, P.T. Collision and grounding mechanics. *The Danish society of Naval Architects and Marine Engineers* (1995): 125–157.

Pitblado, R.M., J. Baik, G.J. Hughes, C. Ferro, and S.J. Shaw. Consequences of liquefied natural gas marine incidents. *Process Safety Progress* 24, 2 (2005): 108–114.

Quy, N.M., J.K. Vrijling, and P.H.A.J.M. Van Gelder. Risk- and simulation-based optimization of channel depths: Entrance channel of Cam Pha Coal Port. *Simulation* 84, 1 (2008): 41–55.

Reason, J.T. *Managing the Risks of Organizational Accidents*, Vol. 6. Ashgate, Aldershot, U.K., 1997.

Smith, L.D., D.C. Sweeney II, and J.F. Campbell. A simulation model to evaluate decision rules for lock operations on the Upper Mississippi River. In *40th Annual Hawaii International Conference on System Sciences, 2007* (*HICSS*'07), Hawaii, IEEE, 2007, p. 56.

Somanathan, S., P. Flynn, and J. Szymanski. The Northwest passage: A simulation. *Transportation Research Part A: Policy and Practice* 43, 2 (2009): 127–135.

Thiers, G.F. and G.K. Janssens. A port simulation model as a permanent decision instrument. *Simulation* 71, 2 (1998): 117–125.

Trading Economics. Qatar exports, 2013a. http://www.tradingeconomics.com/qatar/ (accessed December 10, 2013).

Trading Economics. Qatar imports, 2013b. http://www.tradingeconomics.com/qatar/ (accessed December 10, 2013).

Ulusçu, Ö.S. and T. Altıok. Waiting time approximation in single-class queuing systems with multiple types of interruptions: Modeling congestion at waterways entrances. *Annals of Operations Research* 172, 1 (2009): 291–313.

Ulusçu, Ö.S., B. Özbaş, T. Altıok, and İ. Or. Risk analysis of the vessel traffic in the strait of Istanbul. *Risk Analysis* 29, 10 (2009): 1454–1472.

Van Dorp, J.R., J.R.W. Merrick, J.R. Harrald, T.A. Mazzuchi, and M. Grabowski. A risk management procedure for the Washington State Ferries. *Risk Analysis* 21, 1 (2001): 127–142.

3

Combinatorial Auctions in Turkish Day-Ahead Electricity Markets

KÜRŞAD DERINKUYU,
FEHMI TANRISEVER, FATIH BAYTUGAN,
AND MUSTAFA SEZGIN

Contents

3.1 Introduction

In 1982, Chile reorganized its electricity markets, followed by England and Norway in the 1990s (Hogan, 2002), which speeded up the liberalization politics around the world and also led to fundamental changes to the Turkish markets. Turkish electricity markets have, recently, evolved into a complex competitive business environment with an ever-increasing role of the private sector in production, consumption, and retailing of electricity. Nevertheless, electricity is a unique commodity due to its nonstorability, and hence the demand

and supply of electricity have to be kept in balance in real time, implying a need for a central regulator. In Turkey, Turkish Electricity Transmission Company (TEİAŞ) manages the physical and financial trade of electricity through the Center for Market Financial Settlement (PMUM) and the Directorate of National Load Dispatch (MYTM).

In European electricity markets, power exchanges organize an auction for the determination of the day-ahead market (DAM) prices. These prices established in the DAM are usually accepted as reference prices and affect other electricity markets such as intraday, futures, and forward markets. In the DAM, buyers and sellers submit their bids/offers for each hour of the next day, and the prices are determined to balance supply and demand. DAM price optimization requires solving a combinatorial problem that takes the complex bids of market players and system constraints into account. This problem has to be solved every day within a short amount of time by maximizing the social benefit of the country.

Before giving more details on the DAM, we will explain the alternative electricity market designs to portray the big picture. Then, we provide the information about the historical development of Turkish electricity markets. Later on, we present the general statistics about the current positions of these markets in Turkish electricity production.

3.1.1 Electricity Market Designs

A typical power system can be divided into four layers: generation, transmission, distribution, and trading (retail and wholesale). Generation and trading layers deal with electricity as a commodity, whereas transmission and distribution layers focus on providing electricity services. Electricity markets are designed through the following three main stages (Boisseleau, 2004; Camadan, 2009).

The first stage organizes electricity industry as a whole. In this stage, the policy makers decide on the degree of vertical unbundling, that is, the separation of the market functions traditionally provided by a single utility into functionally independent parts (Sevaioglu, 2013). Hunt (2002) classifies vertical unbundling in four categories:

- *Vertically bundled monopoly*: One company does all the generation, transmission, distribution, and trading activities. Before the liberalization of the sector, this was the main model used almost in every country.
- *Single-buyer model*: In this model, multiple generation companies sell to a single buyer who is also responsible for the transmission, distribution, wholesaling, and retailing activities.
- *Competition on wholesaling level*: This market structure allows the generation companies to sell their electricity through multiple wholesale companies. These wholesalers then sell the electricity to the final consumers.
- *Competition on retailing level*: In this case, even the final consumers have the freedom to choose their electricity providers at the retail level. This level is the highest degree in the unbundling process.

The second stage of the market design is related to the organization of the wholesale markets. It describes the functionality of the markets. Maria (2010) provides advantages and disadvantages of various wholesale market designs as well as their pricing regimes. There are three models considered in this stage:

- *Bilateral agreement model*: Under the absence of a well-defined market organization, the companies make bilateral trade agreements with each other. Since there is no transparency regarding the market prices, this is an inefficient system for all parties.
- *Organized market model*:
  - *Pool model*: The entire electricity trading is transacted via a pooling system. The supply side bids are active in the pool, while the demand side is estimated. The model has an engineering perspective and contains many technical constraints of the supply side. This model is widely used in the United States.
  - *Exchange model*: The participation is not mandatory in this system. Both demand and supply sides are active and could place bids to the market. The model takes an economical perspective and does not contain any technical constraints. This model is widely used in Europe.

- *Mixed model*: Most of the countries combine the bilateral agreement and the organized market models. Bilateral agreements are mostly used for longer-term agreements, and organized markets are used for handling spot market transactions as well as to cover short-term fluctuations in the supply and demand of electricity.

The last stage of the market design is concerned with the financial rules that determine the operations of the market. These rules regulate the contracting intervals, auction types, and pricing mechanisms:

- *Contracting intervals*:
  - *Financial futures and derivatives*: Futures and options to buy and sell electricity are available in most of the developed electricity markets. These derivatives are functional only if they are connected to the physical delivery of electricity (Korkulu, 2008).
  - *Spot market*:
    - *DAM*: Electricity prices of the next day are determined in an hourly basis, 1 day ahead of delivery. These prices are obtained by solving a combinatorial optimization problem. This market provides reference prices to the other markets.
    - *Day-in-market*: Some of the countries allow to continue trading after the closure of the DAM until a few hours before the physical delivery of the electricity. This market enables the participants to revise their positions even after the closure of the DAM.
    - *Balancing market* (*BM*): This is a real-time market to balance the load on the electricity network.
- *Auction types*:
  - *One-sided auctions*: One-sided auctions accept price bids only from the supplier side. This is usually preferred in pool models.
  - *Two-sided auctions*: Two-sided auctions accept bids from both demand and supply sides. This is usually preferred in exchange models.

- *Pricing mechanism*:
  - *Market clearing price model*: This model obtains a clearing price by intersecting the aggregate supply and demand curves. Regardless of the bid prices, all transactions are carried with the same market clearing price.
  - *Pay-as-bid model*: This model determines the accepted bids by intersecting the aggregate supply and demand curves. Then, the suppliers whose bids are accepted receive the price that they ask for their bids. Although the system looks like resulting in lower electricity prices, it causes price inflation since the bidders try to estimate the market price under asymmetric information.
  - *Vickrey model*: In this model, the players of the accepted bids receive the price of the highest bid among the rejected ones. This model is not applicable to two-sided auctions.

Overall, the success of these markets depends on several factors such as transparency, monopoly tightness, participant experiences, and the connection level of each market with physical system.

3.1.2 History of Turkish Electricity Markets

The development of Turkish electricity markets started with the first electricity plant built in Istanbul in 1913. Although the foreign investments were the main driving force for the establishment of electricity plants during these early years, later government policies resulted in the nationalization of the plants in the 1930s. The Ministry of Energy and Natural Resources and Turkish Electricity Authority (TEK) were established in 1963 and 1970, respectively. TEK was formed as a vertically integrated monopoly and controlled all the electricity layers.

With the liberalization policies in the 1980s and 1990s, the private sector gained the rights of building and operating generation plants (Arslan, 2008). The first build–operate–transfer (BOT) plant was built in 1996. Meanwhile, TEK is reorganized into two separate companies: Turkish Electricity Generation and Transmission Company (TEAŞ) responsible for generation, transmission, and wholesaling, and Turkish Electricity Distribution Company (TEDAŞ) responsible for distribution and retailing under the vertical unbundling principles in 1994. Later, TEAŞ is divided into three companies: Electricity Generation Company (EUAŞ) responsible for generation, Turkish Electricity Transmission Company (TEİAŞ) responsible for transmission, and Turkish Electricity Trading and Contracting Company (TETAŞ) responsible for trading in 2001. On the same year, Energy Market Regulatory Authority (EMRA) is established.

After 2001, TEİAŞ acts as both the system operator (of the physical network) and the market operator (for managing financial settlement of the contracts). The restructuring and liberalization of the Turkish market have started in 2003 and gone through four stages until 2014 (Erten, 2006; Erdogdu, 2007; Sitti, 2010):

- *Pre-market age* (*March* 2003–*August* 2006): During this period, there was no organized market for electricity transactions. System operator TEİAŞ manages the electricity grid through EUAŞ by balancing the supply and demand of electricity in real time. Imbalance costs were charged to the participants. After the nationwide blackouts in July of 2006, the market quickly moved to triple pricing age.
- *Triple pricing age* (*August* 2006–*December* 2009): In this period, a day was divided into three intervals: daytime, night, and peak hours. This system failed to establish reference prices in the market and caused the participants to heavily use the BM.
- *Day-ahead planning* (*December* 2009–2011): This new market structure introduced day-ahead hourly prices to trade electricity. In this system, one-sided auctions are held and only the supplier side is allowed to place bids to the market.
- *DAM* (*December* 2011–*now*): Finally, a double-sided blind auction is introduced to the market in which both the supply and demand sides are allowed to place bids for the determination of the day-ahead prices (PMUM, 2012).

On March 14, 2013, the final Electricity Market Law (EMRA, 2013) became active and introduced a new independent company, Energy Markets Management Company (EPİAŞ). The new law resulted in TEİAŞ transferring its market operating responsibilities to this new company. EPİAŞ will become the market operator in the future with a higher stake hold by the private sector. This will organize the market operator as a private entity and function as an independent exchange. In addition, day-in-market design was completed at the end of 2013 and ready to use. However, neither EPİAŞ nor the day-in-market is active as of now, May 2014.

3.1.3 Turkish Electricity Production

Turkish electricity production is evolving rapidly and private sector's share is steadily increasing. Figure 3.1 shows the percentage volume of bilateral transactions in the market between December 2011 and January 2014. The share of public contracts declined from 83% to 72%, while the share of private sector increased by 11% (from 17% to 28%).

Similarly, the role of free consumers who could buy their electricity directly from wholesale companies is increasing. Figure 3.2 shows that their relative size increased from 6% to 33% within 4 years.

Finally, Figure 3.3 provides the DAM, BM, and imbalance order (IO) shares against the total market volume. DAM increased its share from 17% to 34% within 4 years. On the other hand, BM looks steady between 2% and 5%, whereas IO is slightly decreasing from 9% to 7% (ranging between 6% and 11%). As the market gets more mature, we hope to see further declines in IOs.

All the information given in this section is public and could be obtained from PMUM (2014).

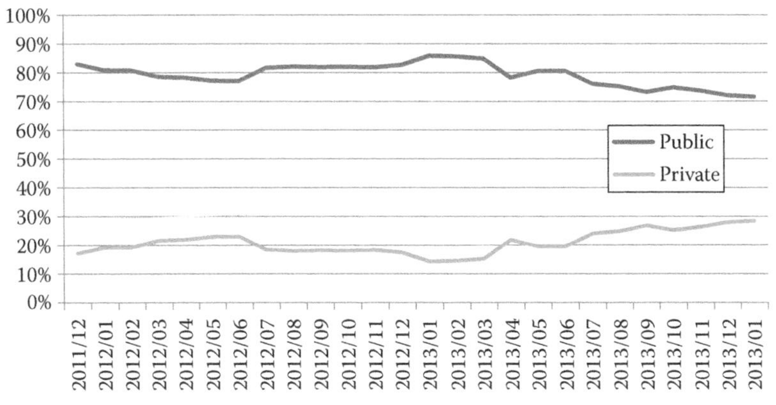

Figure 3.1 Bilateral contract volumes for public and private sectors.

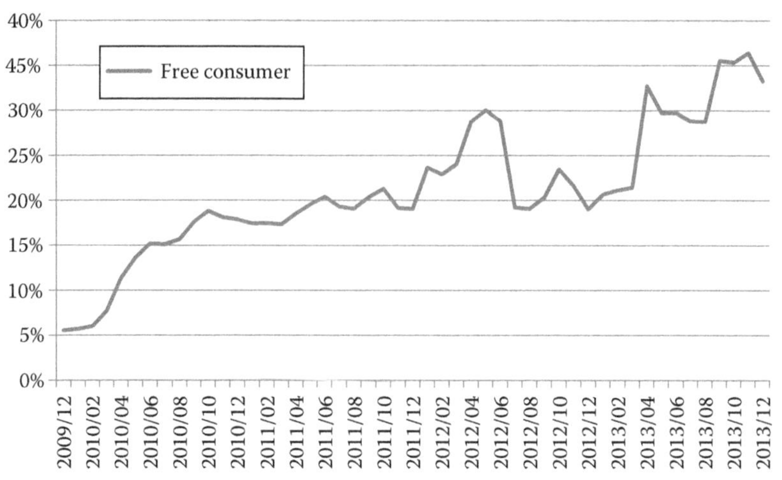

Figure 3.2 Demand of free consumers.

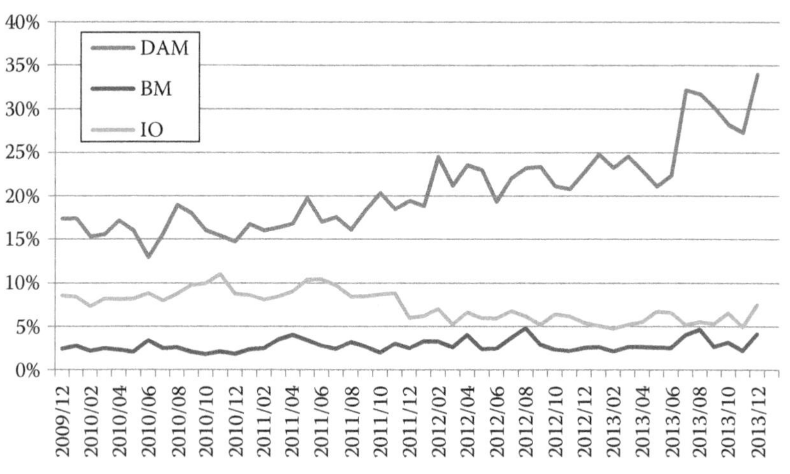

Figure 3.3 DAM, BM, and IO ratios.

3.2 DAMs

Today, power exchanges in European electricity markets organize an auction every day to determine the electricity prices for the delivery of electricity the next day. Most of the auctions are two sided and the exchanges prefer market clearing mechanism as the pricing policy. In these auctions, orders are either accepted or rejected but no side payment is allowed (i.e., linear price regimes). The resulting auction is a combinatorial optimization problem and is hard to solve (Bichpuriya and Soman, 2010).

In Europe, there are independent power exchanges that are responsible for financial markets, and these exchanges run the markets under the physical network constraints. Although traditionally each power exchange controls its own territory, current trend focuses on the market splitting and market coupling. If an exchange generates several local zones due to national boundaries or transmission constraints, it is called market splitting so that each zone may lead to different clearing prices. For instance, the European Power Exchange (EPEX) divides its network into three zones (France, Germany/Austria, and Switzerland) so that different prices could appear in different zones after this market splitting procedure. On the other hand, if several exchange companies act together to increase the efficiency of their system, this is called market coupling. For instance, Nord Pool Spot (NPS) region (Denmark, Estonia, Finland, Lithuania, Norway, and Sweden) uses SESAM coupling system for their markets, whereas Central West European (CWE) region (Belgium, France, Germany, Luxembourg, and the Netherlands) prefers COSMOS coupling system. Moreover, the European Market Coupling Company (EMCC) couples the NPS and the CWE regions. Market coupling considers cross border market coupling bids and allows two or more independent power exchanges to manage the cross border power flow between separate price areas. The reason behind this policy is operating the electricity grid systems more efficiently (Serralles, 2006). This trade may happen either through participating and competing in another zone or merging multiple zones into one complex zone (EMCC, 2014).

In Turkey, PMUM (financial settlement department under TEİAŞ) controls the financial market and settlements. However, it is planned to transfer these rights to EPİAŞ (independent exchange company) in 2014. Turkish electricity market currently has a single zone, but market splitting is scheduled for 2015. Although the Turkish transmission system is physically connected to the European network, there is no planned market coupling in the near future.

In the DAMs, buyers and sellers of electricity place bids (orders) to trade electricity in the following day. Since the buyers and sellers of electricity have special physical requirements for the production and consumption of electricity, their bids in the market reflect these requirements in the DAM. In the following sections, we further explain these bid types and market organization.

3.2.1 Bid Types and Market Organization

In the DAMs, there are three major order types active in most of the European countries: single (hourly) orders, block orders, and flexible orders.

Single orders: Single orders are the buy or sell bids of the participants that are effective only for one time period. The participants submit a set of order price and energy volume for the related time period. Either a stepwise function or interpolation is used to determine accepted volume that corresponds to the clearing price of related period.

Block orders: Block orders are active for several consecutive time periods (at least 4 h in the Turkish market). Volume and price are constant during these time periods. These orders are either accepted or rejected fully, that is, they are not interpolated.

Flexible orders: Flexible orders only have a price–quantity tag and do not provide time period information. Flexible orders are only defined for the supply side, and they are valid for every time period. However, it could be accepted for only one period.

There are also other order types that are not active in any exchange but are already defined in the day-ahead literature (APX, 2014). For instance, *profile block orders* allow submitting different volumes for each hour. This order type is designed particularly for generation units with ramp-up limits. *Partial block orders* allow partial acceptance such as *accept or reject minimum x%*, with x defined by the participant. This product is useful for the generation units with technical minimal limits. *Flexible buy orders* are defined for the demand side of the market. They are designed for participants who have flexibility to change their consumption hour.

3.2.2 Practice and Research on DAM Price Optimization

Given the bids of the market participants, the system operator creates a demand curve and a supply curve and tries to match the supply orders with demand orders while minimizing the price (or maximizing social welfare) for the next 24 h. Research on day-ahead price optimization is relatively new, and there is little academic work on this topic. In the following, we first summarize the academic work on this topic followed by the commercial products used in practice.

3.2.2.1 Academic Research on DAM Optimization The first paper on this topic is published by Meeus et al. (2009). In this paper, authors provide a rigorous MIP formulation of the problem to maximize social welfare, which is defined as the sum of consumer and producer surpluses. The authors also explore the impact of block orders on the complexity of the problem and report how the solution time changes with the density of block orders.

Martin et al. (2014) extend Meeus et al. (2009) by (1) allowing for interpolation of price and quantity bids and (2) considering multiple pricing regions with transmission capacity limits among them. This leads to a more complex MINLP formulation for the day-ahead price optimization. They decompose the model into a mixed integer quadratic master program and a linear pricing subproblem. They also use real data from CWE market. Their method solves the problem to optimality for 60% of the problem instances tested.

Derinkuyu (2014a) introduces an innovative approach to the problem by minimizing prices while representing social welfare in the constraint set. His approach also results in an MIP, which is considerably simpler than Martin et al. (2014). Derinkuyu (2014a) showed that problem size reduction techniques can be effective. He suggests aggregation of single orders and variable elimination. In this approach, single orders are combined to construct one aggregated single order for each time period. The goal behind the aggregation process is to eliminate repeated prices. Since single orders usually contain same prices, there could be more than one binary variable that represents the same price in the same time period. The aggregation process eliminates those unnecessary repetitions and reduces the problem size. Aggregation reduces the problem size (nonzeros in the coefficient matrix) approximately by 60% for the real problem instances.

3.2.2.2 Commercial Products on DAM Optimization The main challenge with the DAM price optimization is the need to solve this problem in a short amount of time. The power exchanges have only 1 h to solve the problem and publicize the market clearing prices. In many power exchanges, the solution algorithm is limited with 15 min to be able to check the correctness of the solution during the leftover time. On the other hand, these problems are hard to solve, and there is no algorithm that can guarantee to find optimal solution within 15 min.

In practice, there are four commercial products used to determine DAM prices: SAPRI, SESAM, COSMOS, and TurkGOPT (under development). The United Kingdom and some African countries use SAPRI, whereas NordPool countries use SESAM algorithm. CWE region designed COSMOS algorithm, and Turkey is currently developing its own algorithm called TurkGOPT. Although each algorithm's objective is to determine day-ahead electricity prices, they may differ in their solution methods as well as flexibility to handle the ever-changing needs of the market participants.

SAPRI is a classical neighborhood search heuristic. SAPRI first finds the equilibrium price considering only single orders in a straightforward manner. Following, it includes all the block orders and initiates an iterative process. The process first tries to exclude the block orders one by one until all block orders satisfy the block order condition. This process described earlier results in a feasible solution. Starting with this feasible solution, the algorithm tries to include some of the excluded block orders one by one to improve solution quality (N2EX, 2013).

SESAM is currently used by the NordPool market. Its development has started in 2004 and it has replaced SAPRI in NordPool as of 2007 (NASDAQOMX, 2014). The major difference of SESAM from SAPRI is its focus on social welfare. In particular, SESAM clearly defines the objective function of the optimization problem as maximizing the social welfare of the market participants (NORDPOOLSPOT, 2014a) and provides a mathematical program. Nevertheless, SESAM also uses a heuristic approach in handling the block orders, and overall SESAM may provide a feasible but not necessarily an optimal solution (NORDPOOLSPOT, 2014b).

COSMOS has developed during the early 2010s and is currently in use in CWE market. COSMOS is a highly functional algorithm for price determination in the DAM. It can handle basic type of orders as well as profile block orders, volume flexible block orders, and exclusive block orders (EPEXSPOT, 2010). In addition, the documentation of COSMOS (EPEXSPOT, 2010) provides a clear mathematical description of the optimization problem as well as the solution method employed to solve it. COSMOS solves a quadratic mixed integer program using a branch and bound approach.

TurkGOPT is being developed by TEİAŞ in collaboration with Turkish universities (Derinkuyu and Tanrisever, 2013). TurkGOPT aims to provide a flexible framework that can handle both social welfare maximization and price minimization as possible objective functions (Derinkuyu, 2014b). TurkGOPT uses variable elimination and aggregation techniques to reduce the problem size. In addition, to handle multiple pricing regions, it suggests the Lagrangian relaxation. Currently, TurkGOPT is designed to handle single, block, and flexible orders, and the algorithm is being improved to include profile blocks, volume flexible blocks, and exclusive block orders as well as buy type of flexible orders.

3.3 Future Trends and Open Problems

As the liberalization process continues, there are many reforms that will be implemented to the Turkish electricity markets (EUD, 2012, 2013). In the future, four types of expansion are expected:

- *Market expansion*: Currently, only the DAM is operational in Turkey. However, in the near future, day-in-market and the physically connected financial markets for trading financial derivatives such as futures and options will become online.
- *Geographical expansion*: Market splitting and market coupling are also considered for the Turkish market. Electricity generation and consumption are geographically dispersed in Turkey, and hence generated electricity needs to be transmitted to the consumption regions over long distances. Therefore, the capacity of the transmission lines is of great importance. Market splitting will help to manage this capacity allocation and pricing problem. Similarly, there are neighboring countries that want to sell or buy electricity to/from Turkey. To manage the electricity network efficiently, there are many opportunities for market coupling.

- *Structural expansion*: Market participants want to get different products such as profile block orders and ramping constraints. These additional flexibilities could be provided in the future.
- *Vertical expansion*: Another emerging issue in Turkey is the establishment of the natural gas market. Since half of the electricity production relies on natural gas, having an organized natural gas market is of significant importance for the market participants in order to better access and manage their risks.

From an academic perspective, there are numerous open research questions (especially in the Turkish market) in this field. Some of them are listed here:

- Although the NP-hardness of the general problem is known, there is no detailed study to show which characteristics of the problem cause the hardness.
- From the regulation perspective, oligarchic behaviors are unknown. How can we figure out illegal actions by examining the market prices and market players' orders? Could we forecast the extraordinary days and prevent it from happening?
- The price effect of renewable resources is not yet studied. Which types of sources are substituted by using renewable ones?
- What is the instability effect of renewable generation units? How could we improve the stability of the network?
- There is a need to analyze the impact of transmission line capacity. What is the effect of transmission capacity lines on the IOs? Which regions need additional transmission capacities?

3.4 Conclusions

This work provides insights into the day-ahead electricity price optimization for the Turkish market. We examine the different market models and indicate the importance of DAMs. Since the problem is complex and has to be solved in a short amount of time, many difficulties arise. Researchers provide several suggestions to overcome these problems. European electricity markets are evolving and this progress brings additional challenges. Literature is still not enough to cover the practical needs of the market, and there are several important open research questions in this field.

References

APX. (2014). Day-ahead auction. apxgroup.com. Retrieved January 15, 2014, from http://www.apxgroup.com/trading-clearing/day-ahead-auction/.

Arslan, S. (2008). Elektrik Enerjisi Sektöründe Serbestleşme, Yeniden Yapılanma, Özelleştirme Uygulamaları ve Dünya Örnekleri. EMRA thesis. Ankara, Turkey.

Bichpuriya, Y., Soman, S. (2010). Electric power exchanges: A review. In *16th National Power Systems Conference*, Hyderabad, India, pp. 115–120.

Boisseleau, F. (2004). *The Role of Power Exchanges for the Creation of a Single European Electricity Market: Market Design and Market Regulation.* Univerisite Paris IX Dauphine, Delft University Press, Delft, the Netherlands.

Camadan, E. (2009). Türkiye Elektrik Piyasası Dengeleme ve Uzlaştırma Mekanizması: Karşılaştırmalı Analiz. EMRA thesis. Ankara, Turkey.

Derinkuyu, K. (2014a). On the determination of day ahead electricity prices. Available at SSRN: http://ssrn.com/abstract=2326435 (September 16, 2013). Submitted to *European Journal of Operational Research.*

Derinkuyu, K. (2014b). Enerji Piyasalarında Bilgi Birikiminin Önemi. *Enerji Panorama*, 11, 50–52.

Derinkuyu, K., Tanrisever, F. (2013). An analysis of the matching algorithms for day-ahead markets in European electricity markets. *Enerji Piyasası Bülteni*, 27, 32–37.

EMCC. (2014). EMCC Optimizer. marketcoupling.com. Retrieved January 15, 2014, from http://www.marketcoupling.com/document/1137/EMCC.

EMRA. (2013). Electricity Market Law, No: 6446. epdk.gov.tr. Retrieved December 17, 2013, from http://epdk.gov.tr/index.php/elektrik-piyasasi/mevzuat?id=143.

EPEXSPOT. (2010). A report for the regulators of the Central West European (CWE) region on the final design of the market coupling solution in the region, by the CWE MC Project. epexspot.com. Retrieved January 15, 2014, from http://static.epexspot.com/document/7616/01_CWE_ATC_MC_project_documentation.pdf.

Erdogdu, E. (2007). Energy market reforms in Turkey: An economic analysis. *Energy Policy*, 35(2), 984–993.

Erten, İ.E. (2006). Rekabetçi Elektrik Dengeleme Piyasalarının Oyun Teorisi Işığında İncelenmesi ve Türkiye Dengeleme ve Uzlaştırma Sisteminin Analizi. EMRA thesis. Ankara, Turkey.

EUD. (2012). Enerji Uzmanlar Derneği. In *Türkiye Enerji Piyasaları için Gelecek Vizyonu Konferansı*. JW Marriott Hotel, Ankara, Turkey, November 20, 2012.

EUD. (2013). Enerji Uzmanları Derneği. In *Strategic Expansion and New Regulatory Perspectives for Turkish Energy Markets*. Meyra Palace Hotel, Ankara, Turkey, December 3, 2013.

Hogan, W.W. (2002). Electricity market restructuring: Reforms of reforms. *Journal of Regulatory Economics*, 21(1), 103–132.

Hunt, S. (2002). *Making Competition Work in Electricity*. John Wiley & Sons, Inc., New York.

Korkulu, Z. (2008). Serbestleştirilmiş Elektrik Piyasalarında Türev Araçların Kullanılması, Vadeli İşlem ve Opsiyon Piyasaları. EMRA thesis. Ankara, Turkey.

Maria, N.S. (2010). Day-ahead electricity market: Proposals to adapt complex conditions in OMEL. Master thesis, Universidad Pontificia Comillas, Madrid, Spain.

Martin, A., Müller, J.C., Pokutta, S. (2014). Strict linear prices in non-convex European day-ahead electricity markets. *Optimization Methods & Software*, 29, 189–221.

Meeus, L., Verhaegen, K., Belmans, R. (2009). Block order restrictions in combinatorial electric energy auctions. *European Journal of Operational Research*, 196, 1202–1206.

N2EX. (2013). N2EX technical overview. n2ex.com. Retrieved January 15, 2014, from https://www.n2ex.com/digitalAssets/85/85812_n2explatformtechnicaloverview27.05.2013.pdf.

NASDAQOMX. (2014). The product group for the financial market. nasdaqomx.com. Retrieved January 15, 2014, from http://www.nasdaqomx.com/digitalAssets/67/67995_financialminutes0106.pdf.

NORDPOOLSPOT. (2014a). Price calculation principles. nordpoolspot.com. Retrieved January 15, 2014, from http://www.nordpoolspot.com/How-does-it-work/Day-ahead-market-Elspot-/Price-calculation/Price -calculation-principles/.

NORDPOOLSPOT. (2014b). Handling of block bids. nordpoolspot.com. Retrieved January 15, 2014, from http://www.nordpoolspot.com/How-does-it-work/Day-ahead-market-Elspot-/Handling-of-block-bids/.

PMUM. (2012). Day Ahead Market. Lecture Series in PMUM, Ankara, Turkey, December 2012.

PMUM. (2014). Genel Raporlar. pmum.gov.tr. Retrieved January 15, 2014, from https://rapor.pmum.gov.tr/rapor/.

Serralles, R.J. (2006). Electric energy restructuring in the European Union: Integration, subsidiarity and the challenge of harmonization. *Energy Policy*, 34, 2542–2551.

Sevaioglu, O. (2013). EE710 Lecture Notes. METU, Ankara, Turkey.

Sitti, K. (2010). Türkiye Organize Toptan Elektrik Piyasalarında Fiyat Olu şumu ve Pazar Gücü Uygulamalarının Tespit Edilmesi. EMRA thesis. Ankara, Turkey.

4

Simulation-Based Inventory Control in a Chemical Industry

BANU Y. EKREN AND M. ARSLAN ORNEK

Contents

4.1 Introduction

In practice, several unexpected factors such as machine failures, human errors, or natural conditions such as weather may cause costly interruptions in production systems. In chemical industries in addition to the aforementioned factors, it is often the case that interruptions occur due to lack of necessary raw materials required for the ongoing production. Although the required raw materials and their amounts are defined in pick lists based on the production type, there are several types of chemicals such as adjust materials and fillers, of which usage and/or amount may not be estimated correctly in advance in those pick lists due to their chemical properties. Hence, lack of these raw materials causes interruptions in the production process and may require long waiting times for their supplements to be replenished from the main warehouse. Because these interruptions and the resulting delays are costly, chemical industries tend to carry these materials as floor stock items (FSIs) at the periphery of the production area to use them when and if required. Therefore, it is significant to determine which materials and in what quantities to carry in the so-called floor stock storage areas (FSSAs) to decrease if not eliminate the number of interruptions. In this chapter, a floor stock problem is solved as an (*s*, *S*) inventory model using simulation optimization.

In the following section, we review and discuss (*s*, *S*) inventory studies and simulation optimization applications from literature. In Section 4.3, we define the problem and propose a solution procedure. In Section 4.4, we analyze and elaborate on our results. Finally, we present our conclusions.

4.2 Literature Review

This chapter considers an inventory control problem where the objective is to determine optimum (*s*, *S*) levels of FSIs using simulation optimization approach; hence, we provide a brief survey from the related literature.

An inventory management system is expected to answer the following three significant questions: (1) how often to review the inventory levels (review period); (2) at what level of inventory to order products (reorder point); and (3) quantity of products to be ordered (order quantity). The literature on these issues is abundant. A comprehensive survey is by Aksoy and Erenguc (1988), which considers multi-item, multilevel products with a joint setup cost structure and coordinated replenishments. The presented studies are mostly solved using heuristic procedures. On the other hand, a general survey on supply chain inventory simulation is provided by Cimino et al. (2010a,b).

Tanrikulu et al. (2010) investigate an inventory problem with multiple items having stochastic demand. The problem is the determination of the replenishment policy that minimizes the total expected ordering, inventory holding, and backordering costs—the so-called stochastic joint replenishment problem. In the study, order setup costs reflect the transportation costs and have a stepwise cost structure where each step corresponds to an additional transportation vehicle.

Kiesmuller (2009) investigates a stochastic multi-item inventory problem where a retailer sells multiple products with stochastic demand and inventory is replenished periodically from a supplier with ample stock. At each order instant, which product and how much to order is decided. They propose a dynamic order-up-to policy where initial order sizes can be reduced as well as enlarged to create full truckloads. They show how to compute the related policy parameters. They also compare their policy with a lower bound and an uncoordinated periodic replenishment policy. As a result, they observe that an excellent cost performance is obtained when the average time between two shipments is not too large and the fixed shipping costs are high.

Roundy and Muckstadt (2000) study base stock policy where an order is placed in each period so as to bring the end-of-period inventory position to a level as close as possible to a given base stock level *s*. They propose a heuristic method for near optimizing base stock policy parameters. Graves (1999) applies the base stock policy for different operative scenarios such as demand pattern variation. Later, Parker and Kapuscinski (2004) implement the base stock policy for two-echelon supply chain. They show that it works well compared to the other inventory policies if the capacity of downstream stages is lower than those of upstream stages. Axsater (2003), on the other hand, considers holding cost minimization under certain fill rate constraints and evaluates the optimal inventory control policy. Tagaros and Vlachos (2001) consider both regular and emergency replenishment policies in their periodic-review inventory control model. The paper shows that considering both regular and emergency replenishment policies works better in terms of costs than a system without emergency replenishment option. Cost minimization is obtained by heuristic algorithms.

The so-called power-of-two (PoT) policy is a widely used heuristic in solving inventory problems where reorder intervals are determined rather than lot sizes (Jackson et al., 1985; Muckstadt and Roundy, 1993). See also Ouenniche and Boctor (2001) and Ekinci and Ornek (2007), for some interesting theoretical discussions on the optimal solution resulted from the PoT policy. The reason this policy is attractive is that it is easier to implement and use in practice but is not necessarily optimal. Jackson et al. (1985) argue that the optimal objective function value, under PoT policy, is within 6% error range for the considered formulation. A PoT policy prescribes for each product i a replenishment interval T_i such that T_i must be a PoT (i.e., 1, 2, 4, 8, …, $2n$, …, where n is a nonnegative integer) times the base planning period, that is, $T_i = 2^n T$. The base period T can be constant or allowed to vary, depending on the model.

Simulation is widely used in identifying the best inventory policies in inventory optimization problems (Tekin and Sabuncuoglu, 2004). This is simply because parameters of the inventory systems are dynamically changing; hence, existing analytical methods are not able to handle all these changes. Also, since enlarged inventory management scenarios are usually complex, they require the use of simulation modeling. A simulation-based approach under these conditions is usually a powerful tool for managing the stochastic behavior of these systems.

Lee et al. (2002) declare the requirement of using simulation not only for inventory management problems but also for analyzing and designing the whole supply chain. A large list of advantages and disadvantages for using simulation approach in supply chain modeling can be found in Ingalls (1998). One of the advantages of using simulation in modeling is that it provides evaluation of multiple performance measures under the effects of different constraints and combinations of critical parameters such as inventory control policies, lead times, demand intensity, and demand variability.

This chapter proposes a solution procedure for managing vendor inventory model, particularly a floor stock control optimization problem of a chemical manufacturing company. We consider a periodic-review inventory control policy for determining the near-optimum levels of reorder and order-up-to levels of FSIs with two replenishment modes: regular and emergency.

4.3 Problem Description and Simulation Modeling

We consider a chemical company operating two production lines (PLs) that produce different types of products. However, the lines may use common type of raw materials. This chapter is motivated because of long interruptions frequently occurring in these PLs due to lack of raw materials during the ongoing production process.

Normally, at the beginning of each shift, raw materials that are required in manufacturing process are available at the PLs. Although the required raw materials and their amounts are defined in the pick lists and they are shipped from the main warehouse to the respective PL, there are several types of chemicals such as adjust materials and fillers, whose usage and/or usage amount may not be estimated correctly in advance due to their chemical properties. In this case, an interruption may occur in that PL and an emergency order is placed for that material type to the main warehouse. This situation usually causes long interruptions in the production process to have these items ready in the PL. These interruptions are very costly and must be reduced if not eliminated. We name the raw materials causing these costly interruptions as FSIs throughout this chapter. So, the objective here is to determine these FSIs and their stock levels to carry at the FSSAs located at the periphery of the PLs. Namely, each PL has its own FSSA with a predefined storage capacity. This way, the company would be able to decrease the number of interruptions in the PLs, which may cause otherwise huge production losses.

The problem is a multi-item, single-echelon (s, S) inventory system where items can be stored in FSSAs that are supplied by the main warehouse with infinite capacity. Here, the two PLs are considered as customers placing orders. Demands for FSIs occur at the beginning of each period. If the demand exceeds the current inventory level at the location, then an interruption occurs. In this case, an emergency order is immediately placed to the main warehouse. The lead times to get the required materials from the main warehouse are assumed to be stochastic and uniformly distributed between 30 and 40 min. We use ARENA 14.0 commercial simulation software to simulate the system.

We consider periodic-review policy (PRP) as the modeling approach in which inventory levels are reviewed after a fixed period of time t_i and an order is placed such that the level of current inventory (I_i) plus the replenishment order quantity (Q_i) equals a prespecified level called the order-up-to-level (S_i) (Equation 4.1) for item i. The review interval t is the time between successive replenishment orders and is constant for a specific item. A PRP is simpler to implement compared to continuous review policies (CRPs) because it does not require the planner to have the capability of monitoring inventory continuously. Suppliers (i.e., the main warehouse) may also prefer PRP because it results in replenishment orders being placed at regular intervals so that transportation cost from the main warehouse is decreased and transportation schedule well organized.

In a multi-item (s_i, S_i) inventory model, the item i is ordered when its inventory level (I_i) decreases to a value equal to or smaller than the reorder level s_i, at time t_i. The order quantity Q_i is such that the inventory is raised to the order-up-to-level S_i:

$$Q_i = \begin{cases} S_i - I_i & \text{if } I_i \leq s_i \\ 0 & \text{if } I_i > s_i \end{cases} \tag{4.1}$$

Pick list items as well as FSIs are kept in the FSSAs. A certain part of the available area of an FSSA is allocated to the FSIs. These areas, items, and their stock keeping units (SKUs) are provided in Table 4.1. We define the area capacities in units of kilogram. For instance, from Table 4.1, we understand that the first FSI's SKU is 225 kg and it is only used by the PL_1. The PL_1 and PL_2 have capacities of 10,800 and 9,000 kg, respectively.

The following issues are considered in the model:

1. First, we determine the raw material types causing interruptions in each PL using the year 2012's 1-year data. We consider 80% of most frequently requested items used in each PL. There are 22 and 18 items in PL_1 and PL_2, respectively (see Table 4.1). Eleven items are common to both lines.

Table 4.1 Standard Packet Sizes (in kg) of Items and Capacities of FSSAs

		PL_1	PL_2			PL_1	PL_2
CAPACITY (KG)		10,800	9,000	CAPACITY (KG)		10,800	9,000
FSI	SKU (KG)			FSI	SKU (KG)		
1	225	X		16	165	X	
2	215	X	X	17	25		X
3	210	X	X	18	25		X
4	210	X	X	19	25	X	
5	240	X	X	20	20	X	
6	180		X	21	25		X
7	215		X	22	20	X	X
8	195	X		23	180	X	
9	1100	X		24	25	X	
10	200	X	X	25	175	X	X
11	200	X	X	26	20	X	
12	190	X	X	27	25	X	X
13	195		X	28	25		X
14	180	X	X	29	25	X	
15	25	X					

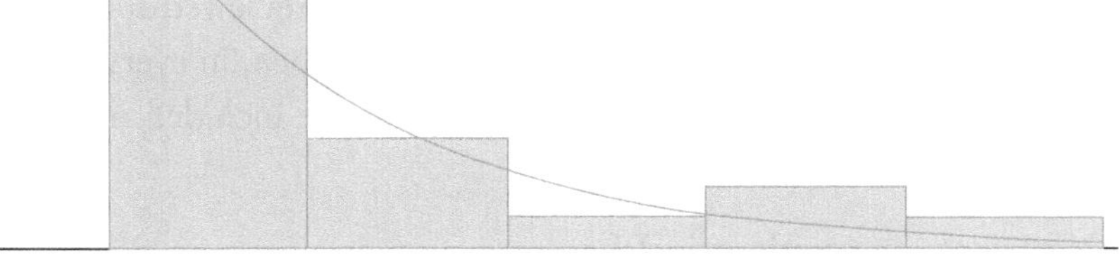

Distribution summary

Distribution:	Exponential
Expression:	1.89e + 003 + EXPO (1.23e + 003)
Square error:	0.012053

Chi-square test
Number of intervals = 3
Degrees of freedom = 1
Test statistic = 1.11
Corresponding p-value = 0.31

Figure 4.1 Input analyzer result for demand distribution of FSI_1 in PL_1.

2. We fit the required amounts of these materials (in kg) to distributions using the input analyzer tool of ARENA 14.0 simulation software. In goodness of fit test, we consider 95% confidence level and check whether $p > 0.05$ or not (Figure 4.1). For a specific item, since we obtain several distributions based on time periods (e.g., days) in a year, we do not provide all of them here. However, we provide the demand distribution of FSI_1 in PL_1 as an example in Figure 4.1. For instance, in that figure based on the χ^2 goodness of fit test, since the p-value is larger than 0.05, we accept it as a good fit.
3. We develop a simulation model of the proposed inventory model.
4. We optimize the model by maximizing review periods (minimizing number of transportations from the main warehouse) satisfying FSSA constraint and a 95% fill rate constraint by considering t_i, s_i, and S_i as decision variables.
5. We consider the so-called PoT policy—2^n—in reviewing items. In this policy, inventory level of an item is reviewed every $t = 2^n$ period of time where n is a positive integer number, for example, 0, 1, 2, ... Also, in our case because a PL has several types of materials to review, by this policy, some material replenishment cycles may overlap because these cycles are

multiples of 2. For instance, let a material be reviewed at every 2^1 (2) h and the other one at every 2^2 (4) h; then, in every two shipments, the second material will also be included in the shipment.

To prevent interruptions in PLs, we assume a minimum fill rate of 95%. Fill rate is defined as a rate at which customer orders can be filled from existing inventory within the parameters of the order request. Hence, in the optimization procedure, we aim to have the required FSI ready in the PL whenever required to prevent any interruption. We use the OptQuest tool in the ARENA 14.0 simulation software for the optimization purpose.

To summarize, the assumptions that are used in the simulation model are summarized as follows:

1. There are two PLs and each has its own FSSA area at the periphery of the line.
2. The FSSA in each PL is limited and the capacity is defined in terms of a kilogram—10,800 and 9,000 kg for PL_1 and PL_2, respectively (see Table 4.1).
3. Eighty percent of the most frequently requested items based on the year 2012's 1-year data for each PL is considered.
4. For each item's demand, amount-fitted distributions are used.
5. Demands arrive at the beginning of each period based on item types.
6. Order-up-to level of item i, S_i, is considered as a multiple of SKU of this item.
7. Reorder level of item i, s_i, is assumed to be a specific percentage (P_i) of its order-up-to-level S_i ($s_i = S_i \times P_i$).
8. For the review period, t_i, PoT policy is considered. We solve each PL independently (first PL_1, then PL_2). The optimum t_i is fixed for the same item in PL_2. Because 11 items are common to both lines, 7 t_i decision variables are considered in PL_2.
9. Items are shipped as SKUs.
10. Simulation is run for 2 years with 3-month warm-up period for 10 independent replications.

In the simulation model, S_i is defined as a multiple of SKU for item i and reorder points—s_i—as a percentage (P_i) of S_i. For instance, if the P_i is 50% in a PL, then $s_i = 0.5 \times S_i$. So, at the review period if the I_i is equal to or less than half of S_i, Q_i is calculated by $S_i - I_i$ divided by this item's SKU weigh. If Q_i is not an integer value then, it is rounded up to the next integer value, and Q_i amount of SKU is shipped from the main warehouse. As an example, in PL_1, item 1's SKU weighs 225 kg. Assume S_1 is 3×225 kg and P_1 is 50%, at the review period, and if I_1 is 200 kg, which is smaller than half of S_1 ($I_1 \leq s_1$; $200 \leq 3 \times 225 \times 50\%$ kg), then Q_1 is calculated as $(337.5 - 200)/225 = 0.61 \approx 1$ SKU of item 1.

In the simulation model, since it is a popular and useful variance reduction technique when two or more alternative configurations are compared, we used common random number (CRN) variance reduction technique. In CRN, the same random number stream is used for all other configurations. Thus, variance reduction is ensured.

The OptQuest optimization tool's efficiency in an (s, S) inventory problem optimization is given by Kleijnen and Wan (2007). This tool combines the metaheuristics of tabu search, neural networks, and scatter search into a single search heuristic (Kleijnen and Wan, 2007). It allows the user to explicitly define integer and linear constraints on the deterministic simulation inputs. It requires the specification of the lower, suggested, and upper values for the variables that are to be optimized. The suggested values determine the starting point for the decision variables, namely, s_i and S_i.

The suggested value choice affects the efficiency and effectiveness of the search. Therefore, first, we near optimize the scenarios and observe the results based on initially selected suggested values. Then, we modify the suggested values by considering results of the first run results. Namely, we specify the suggested values as the near-optimum results of the first run and so narrow the search space.

Recall that the objective function is the maximization of total review periods of items—that is, maximization of total of n_i. We also consider a constraint satisfying maximum total occupied area at a time is smaller than the PL's area capacity. Besides, the fill rate should be minimum 95%. We first solve PL_1, then PL_2. The common items' n_i values are taken fixed while solving the problem for PL_2. That is, the same items will be reviewed at the same periods so that these items will be shipped from the main warehouse at the same time.

4.4 Discussion of Results

Table 4.2 illustrates the OptQuest optimization results. For instance, Table 4.2 indicates that FSI_1 should be reviewed every 128 h and its near-optimum S_1 and s_1 levels are obtained as $4 \times 225 = 900$ kg and $0.65 \times 4 \times 225 = 585$ kg, respectively. FSI_2 should be reviewed every 32 h and the near-optimum S_1 and s_1 levels are $4 \times 215 = 860$ kg and $0.70 \times 4 \times 225 = 602$ kg, respectively, for PL_1. The near-optimum S_1 and s_1 levels are $4 \times 215 = 860$ kg and $0.60 \times 4 \times 225 = 513$ kg, respectively, for PL_2.

All the observations from Table 4.2 are summarized as follows:

- Table 4.2 provides near-optimum order-up-to (S) and reorder (s) levels for each item of two PLs.
- It is observed that the items having small standard packet size (see Table 4.1) usually have larger S_i values than items having larger standard packet size.
- In each PL, the optimum review periods are usually large for items having small standard packet size.
- It seems that there is a positive correlation between review periods and S_i values. Namely, when one increases, the other also tends to increase.

4.5 Conclusions

In this chapter, to the best of our knowledge, it is the first time an inventory model in determining near-optimum level of FSIs in a chemical company is studied. FSIs are raw materials that are commonly and frequently used materials during production. It is significant to have these items in the right amount at the periphery of production environment to prevent interruptions in the manufacturing process. We model the problem as an (s, S) inventory system. Since the items are already carried by the main warehouse in the company, the problem

Table 4.2 OptQuest Results

	PL_1		PL_2			
FSI	s_i (kg)	S_i (kg)	s_i (kg)	S_i (kg)	n_i	t_i (h)
1	0.65	4			7	128
2	0.70	4	0.60	4	5	32
3	0.55	4	0.70	6	8	256
4	0.40	3	0.55	6	7	128
5	0.45	3	0.45	4	6	64
6			0.55	3	8	256
7			0.50	5	8	256
8	0.40	3			6	64
9	0.45	3			6	64
10	0.40	3	0.35	5	4	16
11	0.45	4	0.45	4	5	32
12	0.20	4	0.60	4	4	16
13			0.55	5	9	512
14	0.70	4	0.65	5	5	32
15	0.50	7			5	32
16	0.50	3			6	64
17			0.60	18	6	64
18			0.80	10	3	8
19	0.60	8			3	8
20	0.65	14			3	8
21			0.35	21	6	64
22	0.60	13	0.35	19	4	16
23	0.55	3			5	32
24	0.45	15			4	16
25	0.20	6	0.50	5	3	8
26	0.50	21			6	64
27	0.65	22	0.70	35	5	32
28			0.70	24	5	32
29	0.45	26			2	4

is modeled differently from a classical (s, S) problem. We treat the problem as a multi-item, single-echelon (s, S) inventory system where items can be stored in each of two stocking locations that are supplied by an indoor vendor (i.e., the main warehouse) with infinite capacity. Because each stocking location has a capacity constraint, our aim is to maximize (minimize) review periods (number of shipments from the main warehouse) satisfying required fill rate (95%) and capacity constraints.

References

Aksoy, Y. and Erenguc, S.S. 1988. Multi-item inventory models with coordinated replenishments: A survey. *International Journal of Operations & Production Management* 8(1): 63–73.

Axsäter, S. 2003. Optimal policies for serial inventory systems under fill rate constraints. *Management Science* 49(2): 247–253.

Cimino A., Longo, F., and Mirabelli, G. 2010a. A general simulation framework for supply chain modeling: State of the art and case study. *International Journal of Computer Science Issues* 7(2): 1–9.

Cimino, A., Longo, F., and Mirabelli, G. 2010b. A general framework on products return flows in inventory management: State of the art overview and a supply chain simulator. In *Proceedings of the 16th ISIR International Symposium on Inventories*, August 23–27, Budapest, Hungary, pp. 1–9.

Ekinci, E. and Ornek, A.M. 2007. A heuristic approach for determining lot sizes and schedules using power-of-two policy. *Journal of Applied Mathematics and Decision Sciences* 2007: 18pp., Article ID 53606, doi:10.1155/2007/53606.

Graves, S.C. 1999. A single inventory model for a non-stationary demand process. *Manufacturing & Service Operations Management* 1: 50–61.

Ingalls, R.G. 1998. The value of simulation in modeling supply chain. In *Proceedings of the 1998 Winter Simulation Conference*, Washington, DC, pp. 1371–1375.

Jackson, P., Maxwell, W., and Muckstadt, J., 1985. The joint replenishment problem with a powers-of-two restriction. *IIE Transactions* 17: 25–32.

Kiesmuller, G.P. 2009. A multi-item periodic replenishment policy with full truckloads. *International Journal of Production Economics* 118(1): 275–281.

Kleijnen, J.P.C. and Wan, J. 2007. Optimization of simulated systems: OptQuest and alternatives. *Simulation Modelling Practice and Theory* 15: 354–362.

Lee, Y.H., Cho, M.K., Kim, S.J., and Kim, Y.B. 2002. Supply chain simulation with discrete continuous combined modeling. *Computer & Industrial Engineering* 43: 375–392.

Muckstadt, J.A. and Roundy, R.O. 1993. Analysis of multi-stage production systems. In S. Graves, A.H.G. Rinnooy Kan, and P. Zipkin (eds.). *Handbook in Operations Research and Management Science*, Logistics of Production and Inventory, Vol. 4. North-Holland, Amsterdam, the Netherlands, pp. 59–131.

Ouenniche, J. and Boctor, F.F. 2001. The multi-product, economic lot-sizing problem in flow shops: The powers-of-two heuristic. *Computers & Operations Research* 28(12), 1165–1182.

Parker, R.P. and Kapuscinski, R. 2004. Optimal policies for a capacitated two-echelon inventory system. *Operations Research* 52(5): 739–755.

Roundy, R.O. and Muckstadt, J.A. 2000. Heuristic computation of periodic-review base stock inventory policies. *Management Science* 46(1): 104–109.

Tagaras, G. and Vlachos D. 2001. A periodic review inventory system with emergency replenishments. *Management Science* 47(3): 415–429.

Tanrikulu, M.M., Sen, A., and Alp, O. (2010). A joint replenishment policy with individual control and constant size orders. *International Journal of Production Research* 48(14): 4253–4271(19).

Tekin, E. and Sabuncuoglu, I. 2004. Simulation optimization: A comprehensive review on theory and applications. *IIE Transactions* 36: 1067–1081.

5

Hand Torque Strength in Industry

A Critical Review

MAHMUT EKŞİOĞLU

Contents

5.1 Introduction

Despite technological development and automation efforts, human muscle is still the main source of power for considerable number of industrial jobs and in many daily life activities. Due to this fact, the strength data of the population are essential and fundamental to the design of safe and productive workplaces and equipment and tools in occupational settings as well as the safe and usable design of products

(Norris and Wilson, 1997; Mital and Kumar, 1998; Ekşioğlu, 2004, 2011; Ekşioğlu and Kızılaslan, 2008). The strength evaluations are also necessary for predicting the capability of workers while performing a job requiring strength without incurring injurious strains (Chaffin, 1975; Mital and Kumar, 1998; Ekşioğlu, 2004). Specifically, hand torque strength is an important parameter to be considered for the design and evaluation of a broad range of manual tasks that involve the tightening and loosening of fasteners, threaded parts or connectors, turning knobs, and wheels. In addition, torque strength is also important in some daily life activities such as opening jars, bottles, medicine caps, and turning keys.

Many jobs require high levels of force/torque exertion, and this has a direct impact on safety and performance. High force or torque exertion is one of the main risk factors for the occurrence of musculoskeletal disorders (MSDs: muscle, tendon, and nerve injuries) such as disk herniation, back pain, carpal tunnel syndrome, tendinitis, and DeQuervain's disease, just to name a few. MSDs are among the major problems of occupational safety and health, causing not only human suffering but also heavy cost to industry. Estimated cost of occupational injuries and diseases in the United States in 2007 was $250 billion (Leigh, 2011). In addition, workers working over acceptable and safe work limits will not perform productively for a long time and do a quality work. To set the safe and productive strength limits of tasks, and design safe and usable products, the strength capacities of the workforce and users of the products are needed. Hence, population strength norms should be developed worldwide to determine the reference values for design and evaluation purposes. These norms are also important in clinical settings for evaluation and treatment purposes.

This chapter focuses specifically on the hand torque strength studies. After presenting brief general information of human strength and its importance for industry and daily life, the studies in the reviewed literature are summarized, and important findings and gaps in the hand torque strength data are highlighted so that satisfactory torque strength norms can be developed worldwide. The ultimate aim, of course, is the better application of the torque strength norms to control the risks of MSDs, increase the productivity and quality of work in the workplace, and design safe and usable consumer products for the quality of life.

5.2 Muscular Strength

The limits of human capacity to produce physical work and generate the requisite power are determined mainly by two factors. One of these is the capacity to sustain output over a period of time (endurance) that is a function of aerobic power (i.e., energy expenditure capability). The second is the muscular strength. Muscular strength can affect the design of workplaces and equipment to achieve optimal human performance as well as safety of workers.

Muscular strength can be defined as the capability of a muscle (more correctly a group of muscles) to produce tension (force or torque) between its origin and insertion with voluntary contraction (Gallagher et al., 1998; Kroemer, 2006). (The unit of muscle force is newton [N] or pound-force [lbf] and unit of torque is Nm or lbf-ft). Body segment (hand, elbow, shoulder, back, foot, or some other body part) strength is defined as the force or torque that can be applied by a body segment to an object external to the body (Chaffin et al., 2006). The generated muscular tension is applied to an external object through the skeletal lever system to do work.

Maximum muscular strength is defined as the capacity to produce force or torque with a maximum voluntary muscle contraction. *Voluntary* term here refers to willingness of the subject while performing the physical strength. Thus, measured maximum strength is not the actual maximum of the subject but some lesser value representing what he or she is comfortably expressing at the time with the existing equipment and environmental conditions (Gallagher et al., 1998). The alternative names for muscle strength in the literature are maximum voluntary strength, maximum voluntary exertion, and maximum voluntary contraction (MVC). Maximum muscular force can be exerted for only a few seconds.

Another concept related to muscular strength is the muscular endurance. It is the duration a submaximal force may be held in a fixed position (i.e., static), or the number of times a movement requiring a submaximal force may be repeated (i.e., dynamic).

A muscle's ability to generate force is affected by neural, psychological, physiological, and mechanical factors as well as gender, age, anthropometry, genetics, fitness, training, and motivation. Body posture, joint angle, tool characteristics, and so on are among other factors affecting the strength.

5.2.1 Types of Strength

In general, strength can be divided into two categories: static (isometric) and dynamic. There is also another strength category that is called psychophysical strength. The difference between static and dynamic strength is that dynamic strength involves joint motion whereas static strength does not. Static strength capacity is higher than dynamic strength capacity. Some attempts have been made to predict dynamic strength from static strength data. Table 5.1 provides the basic information of the static and dynamic strength types, and all three types are briefly discussed here.

5.2.1.1 Static (or Isometric) Strength Static strength is defined as the capacity to produce force or torque with a maximum voluntary isometric contraction that results in no movement (or negligible movement) of the object that the muscle is resisting (Kroemer, 1970; Gallagher et al., 1998; Smith, 2006). Body joints are in fixed position

Table 5.1 Types of Strength in Static and Dynamic Activities

ACTIVITY TYPE	STRENGTH TYPE	CONDITIONS	MUSCLE CONTRACTION TYPE
Static exertions (e.g., carry, hold, initiate motions)	Static strength	Fixed postures	Isometric contractions
Dynamic exertions (e.g., lifting, pushing, pulling, repetitive hand motions)	Isoinertial	Body movement with *constant external load*	Concentric or eccentric contractions[a]
	Isokinetic	Body movement with *constant velocity* at specific joint	Concentric or eccentric contractions
	Isotonic	Body movement with *constant muscle tension* (possible only in lab exp.)	Concentric or eccentric contractions

[a] Concentric, shortening; eccentric, lengthening.

during the measurement, and the length of the muscle stays constant during tension. A common example is pushing against a solid wall.

Due to the ease of measurement and control, low equipment cost, short testing time, and test–retest reliability, the assessment of static strength is commonly used. Most of the available strength data are static.

To match the demands of the jobs to the strength capabilities of the workers and product users, static strength norms need to be established. The commonly available static strength data are for grip, pinch, arm, shoulder, torso, and leg strengths.

5.2.1.1.1 Static Strength Measurement Procedure The most popular and scientific method for the measurement of static strength is the procedure proposed by Caldwell et al. (1974). The fundamental instruction is a slow buildup to maximal force in about 1 s and steady exertion sustained for 4 s. The strength datum is the mean score recorded during the first 3 s of the steady exertions. The strength variations during this period should be within ±10% to assume as a valid measurement. If not, the trial must be repeated as many as needed. There should be at least 2 min break between the trials. According to Caldwell protocol, the subject should be informed about the objectives of the experimental study and the procedures before starting the testing session. The instructions should be kept factual and not include emotional appeals. In addition, other factors like rewards, goal setting, competition, and spectators that can stimulate motivation should be avoided. On the other hand, most of the studies available in the literature do not follow this protocol.

5.2.1.2 Dynamic Strength Kroemer et al. (1997) define the dynamic strength as the capacity to produce torque or force by a maximal variometric exertion. It requires joint motion, and thus, it is more complicated to define and measure than static strength measurement. In dynamic activities, muscle length changes, and therefore, the involved body segments move or rotate.

Dynamic strength can be classified as isotonic, isokinetic, and isoinertial contractions (Chaffin et al., 2006). All three can be either concentric (muscle shortens) or eccentric (muscle lengthens). An isotonic contraction internal force within the muscle remains constant while muscle shortens throughout the motion. In isokinetic contraction, velocity of the joint remains constant at the specific joint. In isoinertial contraction, muscle contracts against a constant external load (Table 5.1). Although most of the industrial jobs and human activities are dynamic in nature, due to the measurement difficulty, equipment cost, and time requirement, dynamic strength data are scarce.

The static strength measurement protocol by Caldwell et al. (1974) can be adapted for dynamic strength measurements (Kroemer et al., 1997).

5.2.1.3 Psychophysical Strength In contrary to maximal static and dynamic strength capacities, it does not relate to one-time maximal force exertion capability but the maximum acceptable repetitive force exertion (e.g., lifting, pushing, pulling, gripping, torquing, and so on) capability under prescribed work conditions (e.g., frequency, box size, posture, and so on) for a period of time. It relies on the psychophysical estimation (judgment) of the person. It is time dependent, and estimation is done for over a period of time, usually for 8 h work shift. Psychophysically determined acceptable force/torque/load levels require lab simulation of task of interest, in which the subjects are allowed to adjust the force or load after each attempted performance and continue to make adjustment (for about 30–45 min to simulate an 8 h work shift) until they subjectively believe the force/load to be their maximum they can safely tolerate for the estimated time duration (e.g., Snook, 1985; Ekşioğlu, 2006).

Naturally, the psychophysical strength data values are much smaller than the static and dynamic strength data values. Since they relate to the actual work conditions, they may be considered more realistic and accurate to determine a person's acceptable performance in practical applications. However, the method is time consuming and expensive and requires motivation and cooperation of the subjects to obtain valid results.

5.2.2 Design for Strength

Strength data should be used to guide the design of work, equipment, and products. Force or torque requirement of these obviously should be lower than the maximal strength value. The common approach taken by ergonomists is the design for the weakest member in the specified population. This corresponds usually to the fifth percentile strength (force or torque) value. In addition, to maintain volitional static equilibrium at a joint, moment created by external loads at the joint must be equal or less than the maximum muscle strength moment at the joint. The following force and torque data of the populations are required depending on the design and evaluation: grip, pinch, finger, arm, shoulder, torso, and leg strengths as well as hand torque and push and pull strengths.

5.2.3 Hand Torque Strength

Torque (or moment of force) is the tendency of a force to rotate an object about an axis (torque = force × moment arm). For example, applying force on the handle of a wrench connected to a bolt generates a torque (turning force) that tightens or loosens the bolt. Torque varies throughout joint motion due to change in moment arm and/or change in muscle force production. Not a single muscle but the muscle group interacts as a group to produce torque at a joint. Workers apply force or torque to produce physical work. Mathematically (in both physics and human activity),

$$\text{Work} = \text{Force} \times \text{Displacement} \quad \text{or}$$

$$\text{Work} = \text{Torque} \times \text{Angular displacement}$$

For example, a worker lifting a 10 kg load vertically from floor to a 1 m high shelf does 98 J of work (10 kg × 9.81 m/s^2 × 1 m = 98.1 N m = 98.1 J).

Hand torque strength is important over a broad range of manual tasks because in many industrial work situations and also in daily life, human frequently encounters handle and knob turning tasks such as rotating a door knob to enter a room, opening the lid of a jar or medicine bottle, spinning the knob of any machine, tightening–loosening connectors, or even turning a key. It can be said that, every day, we use our hands and fingers to exert torque and rotational forces in many activities. Therefore, a large number of studies have been conducted by the researchers to model, measure, and predict hand torque strength under specified conditions.

The hand torque strength data available in the literature show variations among the members of each of the studied populations as well as across the populations. Hence, using the strength data of one population for another population may give rise to problems.

5.2.3.1 Types of Hand Torque Torque can be exerted in a number of ways. The orientation as well as the shape of the product (handle/knob/fastener/lid/wheel) is important in torque application. For example, it is easier to apply torque with a T-bar as opposed to a circular knob. Grip force limits the maximal torque exertion for circular knob. With a circular knob, the torque force cannot exceed the tangential force caused by the friction between hand and knob (which increases with increasing grip force) (Daams, 1990, 2006). A small grip force thus limits the maximal force exertion of torque. It is easier to operate a torque device that enables the hand to exert torque without the exertion of much grip force. Hence, torque knobs with levers or handles are recommended for products that require high force and design for weak individuals (Daams, 2006).

1. *Type of hand grasping for torque*: Torque force can be exerted either by whole hand (wraparound grasp) for the case of large knobs or lids or by fingers (fingertip grasp) for the case of smaller knobs or medicine caps.
2. *Maximal and comfortable hand torques*: Torque strength can be determined in two ways: maximal torque values and comfortable torque values. Most of the available torque strength data are maximal type. Only a few studies (e.g., Schoorlemmer and Kanis, 1992) measured comfortable torque. Most of the available torque strength data are static (e.g., Adams and Peterson, 1986; Crawford et al., 2002; Peebles and Norris, 2003; Ekşioğlu and Baştürk, 2013; Ekşioğlu and Recep, 2013).
3. *Clockwise and counterclockwise hand torques*: Torque can be measured in both clockwise and counterclockwise directions. Some researchers found significant difference between them but some others did not. For example, a study by Wieszczyk et al. (2009) reported that the torque generated in counterclockwise direction is greater than that of clockwise. On the other hand, studies, for example, by Rohles et al. (1983), Schoorlemmer and Kanis (1992), and Steenbekkers

(1993), did not find any significant direction effect on torque. Most torque studies involved only one direction. For example, Mital (1986), Ekşioğlu and Recep (2013), and Ekşioğlu and Baştürk (2013) measured clockwise torque only, whereas Berns (1981), Daams (1994), and Imrhan and Loo (1986) measured counterclockwise torque only.

5.2.3.2 Biomechanical Models of Hand Torque Strength Torque can be applied in a number of ways (e.g., screwdriver-like or T- and L-shaped tools [screwdriver, T wrench, and so on], circular [jar lid, knob, and valve wheel, and so on]).

For the case of a circular handle, for instance, to be able to apply torque to open a jar lid, one must generate first enough grip strength to stop slippage (depending on the limiting friction between hand and closure) and second enough wrist-twisting strength to turn the closure. Just to illustrate the case, the biomechanics of torque application for cylindrical handles is expressed as follows:

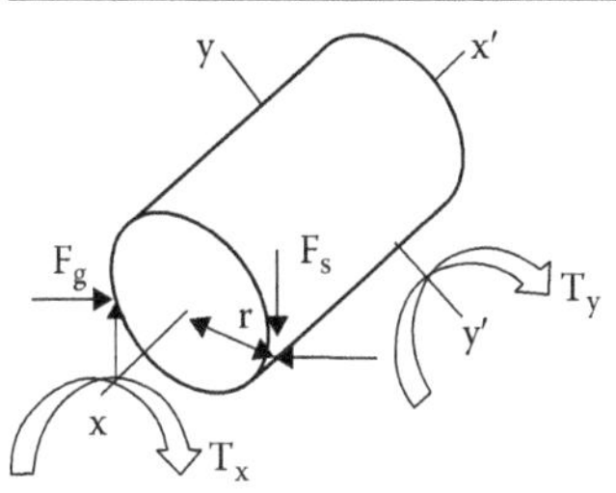

- Torque about the **x–x′** axis of a cylindrical handle: $\mathbf{T_x = F_s r}$.
- *Torque applied by human hand*, for example, to open a circular jar lid: $\mathbf{T_h = F_g \mu r}$ ($=\mathbf{F_s r}$).
  where
  $\mathbf{F_s}$ is the shear force
  $\mathbf{F_g}$ is the hand grip (compressive) force
  $\mathbf{r}$ is the radius of the cylinder
  μ is the limiting friction coefficient between hand and handle
- $\mathbf{T_y}$ (torque about **y–y′** axis) depends mainly on hand torque strength capacity and, within reasonable limits, not on the design of the handle (e.g., T-shaped [T wrench] or L-shaped [lever-type door handle] devices).

5.2.3.3 Factors Affecting Hand Torque Strength It can be stated that all of the factors affecting muscle strength mentioned previously also affect hand torque strength. A combination of some of those and additional factors are listed as follows (mostly adapted from Adams, 2006):

- Orientation to the work surface
- Plane of rotation of hand

- Reach distance
- Type of tool and dimensions of the tool
- Direction of force application
- Resisting force dynamics
- Repetition, duration of single grasp
- Obstructions (barriers to motion)
- Grasp interference
- Grasping method
- Surface roughness or coefficient of friction
- Use of gloves
- Gender
- Age
- Handedness (use of preferred hand)
- Anthropometry
- Fitness and training
- Genetics
- Motivation
- Time of the day

It is almost impossible to investigate the effects of all the factors in one experimental study.

5.3 Hand Torque Strength Studies

A thorough examination of available literature indicated that there have been a considerable number of hand torque studies performed to measure, report, and predict hand torque strength under various specified conditions. The studies generally varied in the subject specifications, procedures used, strength type measured, instruments used, handle type used, and so on. Studies have been conducted in various locations with various sample type, methodology, and instrument. Among those, only a few of them may be considered as normative data to be representative of considered population (e.g., Peebles and Norris, 2003; Ekşioğlu and Baştürk, 2013; Ekşioğlu and Recep, 2013). The reviewed literature on hand torque strength is summarized in the following in chronological order.

Swain et al. (1970) conducted a study to determine the maximum torque that men can apply to knobs in standing position with and without gloves in specific applications.

Method:

- *Subjects*: 120 subjects (96 civilian and 24 military) in two age groups—≤29 years and >29 years
- *Handles*: Diamond knurled knobs with diameters of 0.375 in. (0.95 cm), 0.5 in. (1.25 cm), and 0.75 in. (1.9 cm) and length of 0.5 in. (1.25 cm)
- *Posture*: In standing position, using preferred hand to twist the knob, while steadying the apparatus with the other hand

Results:

- Torque increased with the knob size and decreased with usage of gloves, and side knobs allowed more torque than front knobs.
- Torque strength did not differ significantly between civilian and military personnel.
- Torque strength did not change with age.

In the study by Pheasant and O'Neill (1975), hand–handle linkage was studied to generate useful data for the optimal design of screwdrivers and other devices. Cylindrical handles were chosen for the first stage of the experiments due to their similarity to large number of practical devices use in the real-world activities.

Experiment 1: Maximal steady voluntary torques were exerted by 24 subjects on various handles that were made of polished steel and ranged in diameter from 1 to 7 cm.

Results: Torque strength changed with handle size. For knurled cylinders, 5 cm diameter seems to be optimal.

Experiment 2: 10 subjects were required to exert maximal voluntary torques and maximal thrusts along the axes of the handles.

Results: The actual shape of the handle was unimportant for the forceful activities; however, the effectiveness of the activity was limited by the size of the handle and the quality of the hand–handle interface.

Rohles et al. (1983) conducted an experimental study to determine the wrist-twisting strength capabilities on opening and closing jar lids.

Method:

- *Subjects*:
    - Hundred elderly males and hundred elderly females with age ranging from 62 to 92 years
    - Hundred boys and hundred girls with age ranging from 44 to 58 months
- *Handles*: The lids of eight commercially available food jars with diameters from 2.7 to 12.3 cm.
- *Direction of force*: Both clockwise and counterclockwise.
- *Task*: The strength was exerted on the jar lid snapped on a torquemeter two times in both clockwise and counterclockwise directions.

Results:

- Males were stronger than females in both age groups.
- Age was negatively correlated with torque for older groups; however, in the children, it was positively correlated.
- Torque strength increased with the increase in the diameter of the lid.
- Direction of twisting did not affect torque significantly.
- Grasp and lateral prehension were significant indicators of wrist-twisting strength for all subjects.

Replogle (1983) studied the effect of handle diameter on hand torque strength.

Method:

- *Subjects*: 10 males and 10 females.
- *Handles*: A series of 11 smooth phenolic fiber cylinders ranging in diameter from 0.95 to 8.89 cm.
- *Task*: The participants applied turning force to each cylinder in both clockwise and counterclockwise directions with the preferred hand until the hand slipped.
- *Procedure*: Not mentioned.

Results:

- Female torque capability was about 40% of that of males.
- The grip span (where the fingers and palm just touch without overlapping) and maximum torque diameters did not vary greatly between males and females.
- Torque increased as the square of the handle diameter up to the grip span diameter (2.5 cm). For larger diameters, the torque continued to increase, but at a decreasing rate, and reached a maximum when the diameter is approximately 5 cm.
- The maximum torque was approximately one and one-half times the torque obtainable at the grip span diameter.
- The same diameter handles could be used by males and females to develop maximum torque.

Mital and Sanghavi (1986) examined the effects of several operator- and task-related variables on peak volitional static hand torque exertion capabilities of males and females with common nonpowered hand tools.

Method:

- *Subjects*: A U.S. sample of healthy 55 college student subjects (30 males, 25 females).
- *Tools*: Five different hand tools—two screwdrivers and three wrenches (short screwdriver, 3.7 cm grip diameter and 5.1 cm stem; long screwdriver, 2.9 cm grip diameter and 15.2 cm stem; spanner wrench, 25.4 cm long with 2.2 cm opening; vise grip, 19 cm long with an adjusted opening of 2.2 cm; socket wrench, 24.1 cm long with 1.7 cm opening).
- *Task heights*: Three heights of torque application (eye, shoulder, and elbow height).
- *Postures*: Sit and stand.
- *Reach distances*: Three reach distances (45.7, 58.4, and 71.1 cm) from the seat reference point for the sitting posture and for the standing posture (33, 45.7, and 58.4 cm) from the ankles.
- *Test combinations*: 540 treatment combinations (5 tools × 2 postures × 3 heights × 3 reach distances × 6 angles).
- *Protocol used*: Claimed to be Caldwell et al.'s (1974) protocol but not exactly. Reach maximum in 3 s and hold maximum 1 s.

Peak value was considered maximum. One minute of rest break between test combinations. No repetition.

Results:

- The average female peak torque exertion capability was 66% of the male.
- Type of hand tool, posture, and reach distance were significantly effective on torque strength.
- Both genders generated significantly high torque values with wrenches compared to the screwdrivers. Also higher torque values were generated with the socket wrench compared with the spanner wrench (and the lowest with the vise wrench). The increase in torque was proportional to the increase in the lever arm. Higher torque values were exerted with the screwdriver with the larger grip diameter. The handle diameter of screwdriver but not the length was important for higher hand torque force.
- Both males and females exerted significantly greater torques in the standing posture than in the sitting posture with wrenches, but the opposite was obtained with the screwdrivers.
- The torque exertion capability of both males and females decreased significantly with the reach distance. The maximum torque was exerted at a distance of 33 cm and minimum at a distance of 71.1 cm.
- The effect of the height and angle of torque application, though statistically significant, was not of much practical value for either males or females.
- The isometric shoulder strength appeared to limit the maximum volitional torque exertion capability.

Imrhan and Loo (1986) investigated the container lid variables on maximal counterclockwise (opening) wrist-twisting torque on circular lids in the elderly population.

Method:

- *Subjects*: A U.S. sample of 42 elderly aged 60–97 years.
- *Lid types*: Smooth and rough lid surfaces with four different diameters—31, 55, 74, and 113 mm for each.

- *Posture*: Standing posture except four subjects who preferred to apply force in sitting posture. Preferred handhold the container and the other handhold the tester handle to stabilize it.
- *Procedure*: Not clearly defined. Maximal voluntary opening force was applied to open circular lid containers. Slow buildup allowed and peak force was sustained 2 s.

Results:

- Females were about 75% as strong as males.
- Strength decreased with age and increased with diameter except large smooth lid for which torque decreased.
- Torque increased as diameter increased with rough lids. On the other hand, torque increased with smooth lid up to a certain diameter (74 mm) then decreased.
- Estimated optimum lid diameter was 93 mm.
- Hand breadth, hand length, and hand circumference were all positively correlated with torque.

Nagashima and Konz (1986) examined the effects of jar lid diameter, gripping material, and knurling on torque strength.

Method:

Handles: Diameters of 4.8, 6.7, and 8.6 cm.

Experiment 1:

- *Subjects*: 10 female subjects.
- *Task*: Each subject twisted six jar lids in counterclockwise, which were a smooth and a knurled lid at each of three diameters (48, 67, and 86 mm) with a bare hand, rubber gripper, and a cotton cloth.
- *Procedure*: Not mentioned in detail.

Results:

- The torque strengths increased with increasing lid diameter.
- Torque with rubber gripper was higher than with bare hand and lowest with cotton cloth in the hand.
- No difference between the torques with smooth lid and knurled lid.

Experiment 2: To repeat the first one with a larger, more varied group of subjects to see if knurling was worthwhile.

- *Subjects*: 29 subjects (17 males and 12 females).

Results: Only for 6.7 and 8.6 cm, there were statistically significant differences between smooth and knurled lids.

Adams and Peterson (1986) conducted an experimental study to determine the maximum static hand grip torque that can be exerted during tightening or loosening of circular electrical connectors.

Purpose: Proper design of connectors and task configuration

Method:

- *Factors investigated*: The effects of connector size, grip type employed, orientation of connector, use of work gloves, the reach height of the connectors, and the direction of rotation on hand grip torque.
- *Connector size*: Tightening ring diameters of 2.3, 3.8, and 5.1 cm.
- *Subjects*: 20 males (18–32 years) and 11 females (19–40 years).
- *Procedure*: Modified Caldwell procedure. Three-second static force was exerted by each of the subjects in standing. Torque was applied to simulated connector rings.

Results:

- Hand grip torque increased by connector diameter.
- The orientation of connector was effective on torque strength.
- The usage of gloves also resulted in slightly higher torque values.
- Height and direction of rotation had little effect on torque strength.
- Males were significantly stronger than females in both tightening and loosening.

Imrhan and Jenkins (1999) investigated the effects of surface finish, wrist action, arm position, and hand laterality on wrist flexion and extension torque capabilities of male and female adults in simulated maintenance tasks.

Method:

- *Handles*: Knurled and smooth two identical solid cylindrical aluminum handles (diameter, 5.72 cm; length, 13.97 cm).
- *Sample*: Healthy 10 males (manual workers: 28–43 years) and 10 females (1 manual worker and 9 homemakers: 25–40 years).
- *Task*: Generation of maximal volitional static torques on a cylindrical handle snapped on a portable torquemeter over 24 different test conditions.
- *Procedure*: Subjects gripped handles bare-handed with power grip in comfortable standing with their arms fully extended and approximately in sagittal planes. Peak force was recorded. Rest time was ≥1.5 min between exertions. No further details were provided.

Results:

- Overall, males were twice as strong as females. However, the sex difference depended on certain task and handle variables, such as handle surface and wrist action.
- In both males and females, torque was greater in extension than in flexion and with the knurled handle compared to the smooth one.

Kim and Kim (2000) studied the effects of body posture and of different types of common nonpowered hand tools on maximum static volitional torque exertion capabilities of Korean people.

Method:

- *Hand tools*: Screwdriver, socket wrench, cylindrical handle, rotating knob, steering wheel.
- *Subjects*: A Korean sample of 15 healthy male and 15 female university students.
- *Experimental variables*: 15 body postures and 5 hand tools with a total of 75 test combinations/subject.
- *Procedure*: Modified Caldwell et al.'s (1974) (though not mentioned). Subjects were asked to build up to the maximum torque gradually, without jerking, over 3 s period, and then hold it at the maximum for about 1 s. Repetition at least twice

to be within ±10% difference. Higher value was chosen as MVC. Rest breaks were ≥1 min between trials.

Results:

- The torque exertion capability was significantly affected by the type of tool and posture for both males and females: relatively higher torques were exerted in the order of steering wheel, wrench, handle, knob, and screwdriver. It may be said that torque exertion was affected by lever arm of the tools.
- Female torque strength was about 51.5% of males.
- Both males and females exerted the most torque in standing, eye height, and tool axis horizontal, whereas males showed the least value in standing, overhead, and tool axis vertical and females exerted the least torque in kneeling on one knee, overhead, and tool axis vertical.
- Higher torques were exerted at shoulder and eye height, whereas relatively lower torques were found at elbow and overhead height.

The study by Voorbij and Steenbekkers (2002) aimed to answer the question of what maximum torque should be allowed for opening a jar (opening torque).

Method:

- *Opening torque*: Measured in people over 50 years old, with a jar lid. The jar, which was made of aluminum, weighed 650 g. The lid had a diameter of 66 mm, while at its widest point, the jar was 75 mm in diameter. The total height was 113.5 mm.
- *Subjects*: A Dutch sample of healthy 750 subjects: 123 of them were aged between 20 and 30 years as a reference group, and 627 were over 50 years of age.
- *Posture*: The wrist-twisting force was measured while the subject was standing. The subject was asked to adopt the posture normally used for opening jars. One hand was on the lid while the other grasped the jar.
- *Procedure*: The subject was instructed to build up force to maximum and to hold this maximum force until the second

attendant called a stop. This attendant checked for an acceptable length of the constant phase in the force graph (i.e., 1 s). The force exertion was repeated once after a 2 min period of rest. Measurement was made twice, and the peak value of the two was taken as the maximum torque value.

Results:

- The preferred way of opening a jar was with both hands: one on the lid and one on the jar.
- Laterality was significant: the preferred hand grasped the jar.
- The required torque for opening a jar should not exceed 2 N m to accommodate >95% of users between 50 and 94 years.

Crawford et al. (2002) investigated the impact of shape, diameter, and height of lid on static wrist torque and also examined opening torque of commercially available food jars. (Grip and pinch strengths were also measured.)

Method:

- *Subjects*: 40 healthy adults: 20 young (10 males and 10 females with age 20–39 years) and 20 older with age 69–81 years for the males and 60–72 years for the females.
- *Test pieces and combinations*: 12 nylon test pieces with 10, 20, and 30 mm in length (9 circular with 20, 50, and 80 mm diam. and 350 mm diam. square with rounded edges). Each participant made two maximal wrist torque exertions for each piece of six test pieces.
- *Body posture and torque direction*: The torque exertions were made in a *standing position* turning the test piece in a counterclockwise direction (to open a jar) using preferred hand to create the torque and the nonpreferred hand to hold the circular fixing point.
- *Opening forces* (*required to open a variety of food products*): Measurements were carried out by placing each jar in the fixture and opening the product with the preferred hand.
- *Torque direction*: Counterclockwise with preferred hand.
- *Procedure*: Not reported.

Results:

- Higher torques were generated on lids that were square compared to those that are circular of the same diameter.
- As lid diameter and lid height increased, torque increased for test pieces between 20 and 50 mm diameters.
- A linear relationship for torque existed for the test pieces between 20 mm diameter and 10 mm height and 50 mm diameter and 30 mm height.
- Height, weight, hand length, and hand breadth were positively correlated with torque strength.
- The lid surface area and torque strength were highly correlated; thus, a linear model was developed to describe this relationship:

  $$\text{Torque} = -7.26 + (1.23 \times \ln \text{surface area})$$

  For the weakest group of participants, the model was described by

  $$\text{Torque} = -5.69 + (0.94 \times \ln \text{surface area})$$

  The model could be used to predict maximal torque closure levels for use in the packaging industry.

Peebles and Norris (2003), in their normative static strength study, measured six strength capabilities in freestanding postures: finger-push strength, pinch–pull strength, hand grip strength, push and pull strength, and the hand torque strengths (wrist-twisting and opening). For our purpose, hand torque strengths only were discussed here.

Method:

- *Wrist-twisting strength*: Twisting force was exerted with dominant hand in a clockwise direction on a variety of knobs and handles: (1) door lever (diameter 15 mm, length 170 mm), (2) door knob (diameter 65 mm, depth 45 mm), (3) circular knob (diameter 40 mm, depth 20 mm), (4) ridged knob (length 40 mm, depth 15 mm), (5) butterfly nut (length 40 mm, depth 10 mm), and (6) tap (diameter 50 mm, depth 40 mm).

All handles were positioned at elbow height and orientated vertically (vertical wrist-twisting strength). The ridged knob, butterfly nut, and tap were also orientated horizontally (horizontal wrist-twisting strength).

- *Opening strength*: Measured on three custom-made aluminum jars (height 125 mm) with smooth and knurled lids (diameters 45, 65, and 85 mm). The jar was held with one (preferred) hand, and a static twisting force was exerted with the other hand on the lid of the jar.
- *Subjects*: A UK sample of healthy 150 males and females aged from 2 to 86 years.
- *Procedure*: Caldwell et al.'s (1974) with modification (not the middle 3 s steady-state average value but the peak value is considered). The higher of the two repetitions was taken as the maximum.

Results:

- All torque strength data exhibited a normal distribution.
- Maximum strength increased throughout childhood, peaked in adulthood, and then decreased with age from around 50 years.
- Each successive age group (2–5, 6–10, 11–15 years) was found to be significantly stronger than the previous for all measurements. Generally, however, no significant differences were found within the adult (16–20, 21–30, 31–50 years) or older adult (51–60, 61–70, 71–80, 81–90 years) age groups. No significant differences in maximum strength were generally found between 11–15 and 60–80 years old or 6–10 and 80–90 years old.
- Female/male strength ratio ranged from 55% to 75%. However, no significant differences in maximum strength were found between male and female children.
- No significant correlations were found between the six strength measurements.
- The handle or control type, the direction of force, and the number of hands used all significantly affected the amount of force that could be exerted.

- Significant difference between two relatively similar measurements—wrist-twisting strength (on a variety of handles placed at elbow height) and opening strength on jars.
- Knurling on lids increased the opening strength.

Nayak and Queiroga (2004) studied the pinch grip, power grip, and wrist-twisting strengths of healthy older adults in the United Kingdom.

Method:

- *Subjects*: 150 subjects (all Caucasian origin), 65 males and 85 females, in the age range of 55–85 years.
- *Equipment*: A torquemeter to measure wrist-twisting strength. The main body of the unit was 90 mm long with a diameter of 44 mm. At one end of the unit, a 50 mm diameter plastic lid was fixed to which a torque was applied. The lid thickness was 10 mm with a slightly rough texture.
- *Posture*: Seated that followed the guideline by the American Society of Hand Therapists.
- *Procedure*: Subjects, while seated, were instructed to hold the jar-shaped unit with the nonpreferred hand (power grip position) and to apply the twisting movement on the lid with the preferred hand (spherical grip position). They were instructed to exert their maximum possible torque and to hold it for about 5 s.

Results: A removal torque of 1.3 N m could be used as a guide for the design of screw tops for child-resistant bottles, such as medication containers.

Miller et al. (2005) used a simple device to quantify twisting strength necessary to perform daily activities.

Method:

- *Subjects*: 64 normal subjects (46 females and 18 males) and 13 arthritic patients (9 females and 4 males) with arthritis of the thumb carpometacarpal joint.
- *Handles*: Five disks of 8 mm thickness and diameters of 2.5, 5, 7.5, 10, and 12.5 cm were fabricated from plastic, the edges were rounded and smoothed, and each disk was rubber coated.

- *Task*: Apply a twisting force to each of the five disks with each hand, in both the clockwise and counterclockwise directions, for three trails of each.
- *Posture*: Each subject adopted the standard testing position and applied a twisting force to each of the five disks.
- *Measurement protocol*: Maximum values were recorded for each trial, and the results of the three trials were averaged.

Results:

- Males applied greater torques than females.
- The dominant hand applied greater torque.
- Subjects diagnosed with carpometacarpal arthritis could not apply normal levels of torque.
- There were no differences in the test–retest group due to the day of testing.
- All disk sizes generated significantly different torques.
- Ulnar and radial torques were similar.
- The torque values recorded were lower than those reported by Voorbij (2002), Crawford et al. (2002), and Peebles and Norris (2003).

The study by Yoxall et al. (2006) investigated the forces applied by consumers to open a jar lid, specifically to opening a wide-mouth vacuum lug closure (such as those used for jams, sauces, and pickles).

Method:

- *Subjects*: 235—138 males (aged from 8 to 93) and 97 females (aged from 8 to 95).
- *Handle*: Jar lid with a diameter of 75 mm.
- *Task*: Open the jar as normally done by the users. Subject could pick up the jar or leave it resting on the table. No type of grip was suggested, and multiple attempts using different postures or grips were allowed.
- *Torque direction*: Counterclockwise.
- *Response*: Peak applied torque.
- *Posture*: As preferred, sit, or stand.

Results:

- Most of the females would struggle to open some jars.
- Males were considerably stronger than females, so most of them would not have trouble to open the bulk of jars.
- After 60 years old, strength started to decrease rapidly.
- For a 75 mm jar, 15% of women of any age would struggle with 50% of the jars they bought, indicating that openability of jars of this type was a significant problem.

The study by Kong et al. (2007) investigated the effects of screwdriver handle shape, surface material and workpiece orientation on torque performance, total finger force, and muscle activity in a maximum screwdriver tightening torque exertion task.

Method:

- *Handles*: 24 screwdriver handles, each with a length of 130 mm, were constructed with factorial combinations of longitudinal cross-sectional shape (circular, hexagonal, triangular), lateral cross-sectional shape (cylindrical, double frustum, cone, reverse double frustum), and surface material (plastic, rubber coated). The nominal diameter of all the handles in cross section was 45.0 mm, but the maximum dimension in cross section depended on the handle's cross-sectional shape.
- *Material*: rubber and plastic.
- *Subjects*: 12 healthy male university students.
- *Task*: Perform maximum clockwise screw-tightening exertions using screwdriver handles with 3 longitudinal shapes (circular, hexagonal, and triangular), 4 lateral shapes (cylindrical, double frustum, cone and reversed double frustum), and 2 surfaces (rubber and plastic) (all 24 screwdriver handles). Six of the 12 subjects tested maximum torque exertion from the vertical orientation; the other 6 subjects exerted their maximum torque from the horizontal orientation.
- *Procedure*: Caldwell procedure was used (though not mentioned). The dominant hand maximum torque exertion to the screw assembly on the torque sensor in standing posture with a straight elbow in the horizontal workpiece orientation

or approximately 90° elbow flexion in the vertical workpiece orientation. Exertion duration was 4 s and was repeated two times for each handle, with 2 min of rest time between trials. Steady-state average value was used as the maximal torque strength.

Results:

- Torque output with rubber handles was 15% greater than plastic handles.
- The vertical workpiece orientation was associated with higher torque output (5.9 N m) than the horizontal orientation (4.69 N m).
- Screwdrivers designed with a circular or hexagonal cross-sectional shape resulted in greater torque outputs (5.49, 5.57 N m), with less total finger force (95, 105 N).
- Reversed double-frustum handles were associated with less torque output (5.23 N m) than the double-frustum (5.44 N m) and cone (5.37 N m) handles.
- Screwdriver handles designed with combinations of circular or hexagonal cross-sectional shapes with double-frustum and cone lateral shapes were optimal.

Seo et al. (2007) conducted an experimental study to develop a model to describe the relationship between grip and torque.

Method:

- *Subjects*: 12 subjects (6 females and 6 males) with age ranging from 21 to 35.
- *Handles*: Diameters of 45.1, 57.8, and 83.2 mm.
- *Task*: Grasp the cylindrical object with diameters of 45.1, 57.8, and 83.2 mm in a power grip, and perform maximum torque exertions about the long axis of the handle in two directions: the direction the thumb points and the direction the fingertips point. Maximum torque, grip force, total normal force, and fingertip/thumb force were measured.
- *Posture*: Seated with elbow flexed about 90° and forearm horizontal and grasped a vertical cylindrical handle with the right hand in a power grip.

Results:

- Handle diameter had a significant effect on torque exerted.
- Hand torque was greater when the torque on a cylinder was applied in the direction the fingertips point.

Seo et al. (2008) investigated the relationship among friction, applied torque, and axial push force on cylindrical handles.

Methods:

- *Subjects*: 12 healthy participants.
- *Handles*: Handle diameters—5.78 and 5.1.2 cm for the rubber and aluminum handles.
- *Task*: Exert anteriorly directed maximum push forces along the long axis of aluminum and rubber handles while applying deliberate inward or outward torques, no torque (straight), and an unspecified (preferred) torque.
- *Posture*: Standing.
- *Protocol*: All data were averaged over 2 s during maximum exertions. No further information is provided.

Results:

- Axial push force was 12% greater for the rubber handle than for the aluminum handle.
- Participants exerted mean torques of 1.1, 0.3, 2.5, and -2.0 N m and axial push forces of 94, 85, 75, and 65 N for the preferred, straight, inward, and outward trials, respectively. Left to decide for themselves, participants tended to apply inward torques, which were associated with increased axial push forces.
- Participants appeared to intuitively know that the application of an inward torque would improve their maximum axial push force.

Wieszczyk et al. (2009) aimed to determine the effect of height of hand wheel of an industrial valve on the maximum torque production and risk of injury to the shoulders and back of workers.

Method:

- *Valve wheel*: 45 cm in diam.
- *Task*: Maximum torque exertions in the clockwise and counterclockwise directions at three heights (knee, chest, and overhead) while standing.
- *Subjects*: 24 healthy power plant mechanics or operators (23 males and 1 female: 32–61 years).
- *Procedure*: Participants exerted two maximum torques for each condition of height and direction, with at least 2 min of rest between consecutive torque exertions. Participants wore leather gloves during all torque exertions. Maximum torque was the average of the two trials. No further details were given. Caldwell protocol possibly was partially implemented.
- *Torque direction*: Both in clockwise and in counterclockwise.

Results:

- Torque generated in the counterclockwise direction was greater than that of clockwise.
- Ten percentage or greater torque was exerted at the overhead level than at the chest level. However, there was no difference in maximum torque between knee and overhead levels and between knee and chest levels.
- Design engineers should avoid placing hand wheel valves at knee height or lower.

Rowson and Yoxall (2011) conducted a study to determine the effect of different hand grips on maximum opening torque.

Method:

- *Subjects*: 34 (19 females and 15 males).
- *Task*: Apply twisting force to open the jar with 7 different gripping type and 3 different jar diameters of closure (21 tests for each participant).

Results:

- Female participants generally produced lower torques than males.

- Different grip styles were then seen to produce different peak torque values.
- Only a limited number of grip styles applied by women gave them a sufficient strength to be able to open the jar.
- The spherical grip choice produced the highest torque for the females, and they are likely to use a spherical grip on containers of this type.
- In males, all of the grip styles produced maximum torques above the torque required for opening jars.

The study by Ekşioğlu and Recep (2013) aimed to establish the static hand torque strength norms of healthy adult female population of Turkey and to investigate the effects of handle type, posture, age group, job group, and several anthropometric variables on hand torque strength.

Method:

- *Subjects*: 257 females (18–69 years).
- *Handles*: Cylindrical (diam., 51 mm; length, 113 mm), circular (diam.: 60 mm), ellipsoid (with axis lengths: 55.6 and 42 mm), and key.
- *Task*: Maximum voluntary static torque strengths of dominant hand were measured both in sitting and in standing with four types of handles.
- *Procedure*: Caldwell et al.'s (1974) was used.
- *Posture*: Both freestanding body and neutral sitting posture with the shoulder, elbow, and wrist about in neutral posture.
- *Direction*: Clockwise.

Results:

- Torque strength norms were developed for the adult females (18–69 years).
- The handle type, age group, and job group significantly affected torque strength. The highest values were obtained with cylindrical handle followed by circular, ellipsoid, and the lowest with key handle.
- The torque strength peaked in 30–39 age group for nonmanual and in 40–49 age group for manual workers.

- Manual workers were stronger than nonmanual workers.
- Marginally higher strength values were recorded in standing posture.
- Overweight group was marginally stronger than normal weight group.
- Grip strength and some of the anthropometric variables, such as forearm circumference and hand breadth, were positively correlated with torque strength.
- The comparison results showed similarities and differences with some other nationalities.

Ekşioğlu and Baştürk (2013) estimated the static hand torque strength norms of healthy adult male population of Turkey and investigated the effects of handle type, posture, age group, job group, and several anthropometric variables on hand torque strength.

Method:

- *Subjects*: 257 males (18–69 years).
- *Handles*: Cylindrical (diam., 51 mm; length, 113 mm), circular (diam.: 60 mm), ellipsoid (with axis lengths: 55.6 and 42 mm), and key.
- *Task*: Maximum voluntary static torque strengths of dominant hand were measured both in sitting and in standing with four types of handles.
- *Procedure*: Caldwell et al.'s (1974) was used.
- *Posture*: Both freestanding body and neutral sitting posture with the shoulder, elbow, and wrist about in neutral posture.
- *Direction*: Clockwise.

Results:

- Torque strength norms were developed for the adult males (18–69 years).
- The handle type, age group, and job group significantly affected torque strength. The highest values were obtained with cylindrical handle followed by circular, ellipsoid, and the lowest with key handle.
- The hand torque strength peaked in 40–49 age group for both manual and nonmanual job groups in the three handles

(ellipsoid, circular, and key). On the other hand, for cylindrical handle, hand torque strength peaked in 18–29 age group for both manual and nonmanual job groups.
- Manual workers were stronger than nonmanual workers.
- Marginally higher strength values were recorded in standing posture.
- Grip strength, height, hand length, hand breadth, and forearm circumference were positively correlated with hand torque strength.
- Body mass index did not have significant effect on torque strength considering only normal and overweight.
- The comparison results showed similarities and differences with some other nationalities.

5.3.1 Hand Strength Data

A summary of hand torque strength data of the world populations obtained from some of the aforementioned summarized studies is given in Table 5.2. As it can be seen, torque values show some cross-national variations. Some of the variations can be attributed to the experimental conditions, sample size, instrument and methodology used, and age range of the subjects studied. Part of the variation, however, may be due to the differences among the characteristics of the nations.

5.3.2 Summary and Critics of Findings

1. The following can be driven from the reviewed studies in terms of methodology:
   a. All studies reviewed measured static hand torque strength. None was about dynamic and psychophysical hand torque strengths.
   b. Only a few of the studies may be considered normative. The rest of the studies tried to identify the effects of some factors on the hand torque strength with very small sample sizes.
   c. Most of the studies were about young and middle-aged healthy adult population; only a few were about elderly and children.

Table 5.2 Summary Results of Some Torque Strength Studies in the Literature

STUDY	COUNTRY	SAMPLE SIZE AND TYPE	AGE (YEARS)	HANDLE/OBJECT	DIAMETER	TORQUE STRENGTH (Nm)	
						MALE	FEMALE
Nagashima and Konz (1986)	United States	Exp. 1: 10 F	NM[a]	Jar lid	48 mm		3.17
					67 mm		5.02
					86 mm		6.04
		Exp. 2: 17 M and 12 F			67 mm	Rubber Bare	Cloth
						Smooth:	
						9.8 7.8	6.3
						Knurled:	
						8.9 7.9	6.3
					86 mm	Smooth:	
						11.3 10.4	8.7
						Knurled:	
						10.9 10.1	9.5
Imrhan and Loo (1986)	United States	42 M and F	60–97	Rough and smooth lid with diameters		Rough	Smooth
					113 mm	5.01	3.29
					74 mm	4.20	4.19
					55 mm	3.30	3.25
					31 mm	1.62	1.53
Imrhan and Jenkins (1999)	United States	10 M and 10 F	28–43	Cylindrical handle	57.2 mm	9.11 ± 0.72	4.68 ± 0.39
Kim and Kim (2000)	Korea	15 M and 15 F	18–29	Cylindrical handle	34 mm	11.4 ± 1.52	5.96 ± 2.23
						12.66 ± 1.73	6.94 ± 2.08

(*Continued*)

Table 5.2 (*Continued*) Summary Results of Some Torque Strength Studies in the Literature

STUDY	COUNTRY	SAMPLE SIZE AND TYPE	AGE (YEARS)	HANDLE/OBJECT	DIAMETER	TORQUE STRENGTH (Nm)	
						MALE	FEMALE
Voorbij and Steenbekkers (2002)	Netherlands	750 M and F	20–30	Jar lid	66 mm	8.7 ± 2.2	5.6 ± 1.4
			50–54			7.6 ± 1.8	4.8 ± 1.5
			55–59			7.6 ± 2.3	4.7 ± 1.4
			60–64			6.4 ± 1.8	4.8 ± 1.4
			65–69			6.5 ± 2.1	4 ± 1.2
			70–74			5.4 ± 2.1	3.7 ± 1.1
			75–79			5 ± 1.7	3.5 ± 1.3
			80+			4.9 ± 1.7	3.4 ± 0.9
Peebles and Norris (2003)[b]	United Kingdom	150 M and F	21–30	Disk	4 cm	4.1 ± 1.8	3.5 ± 1.3
						4.5 ± 1.7	3 ± 1.3
			31–50			4.2 ± 1.1	3.5 ± 0.6
						3.2 ± 1.4	2.6 ± 0.6
			51–60	Butterfly nut		3.9 ± 1	2.4 ± 0.4
						4.3 ± 1.6	2.8 ± 0.7
			61–70			3.6 ± 0.8	2.7 ± 0.5
						3.2 ± 0.5	2.3 ± 0.6
Miller et al. (2005)	United States	46 F and 18 M and 13 arthritic patients	19–74	Disk	5 cm	2.16 ± 0.63	1.44 ± 0.41
					7.5 cm	3.37 ± 0.86	2.20 ± 0.63

Seo et al. (2008)	United States	6 M and 6 F	21–35	Cylindrical handle (rubber)	57.8 mm	Inward torque 8.7 ± 2.5	Inward torque 3.5 ± 2.1
				Cylindrical handle (aluminum)		Inward torque 6.9 ± 1.3	Inward torque 2.8 ± 1.7
Ekşioğlu and Recep (2013)	Turkey	257 F	18–69	Ellipsoid handle	55.58 mm		Sit:2.74 ± 0.71 Stand: 2.96 ± 0.78
				Circular handle	59.98 mm		Sit: 3.30 ± 0.86 Stand: 3.51 ± 0.88
				Key handle			Sit: 1.59 ± 0.39 Stand: 1.65 ± 0.42
				Cylindrical handle	50.7 mm		Sit: 5 ± 1.3 Stand: 5.3 ± 1.38
Ekşioğlu and Baştürk (2013)	Turkey	257 M	18–69	Ellipsoid handle	55.58 mm		Sit: 4.12 ± 1.09 Stand: 4.25 ± 1.12
				Circular handle	59.98 mm		Sit: 4.83 ± 1.12 Stand: 5.10 ± 1.28
				Key handle			Sit: 1.97 ± 0.38 Stand: 2.05 ± 0.45
				Cylindrical handle	50.7 mm		Sit: 6.87 ± 1.82 Stand: 7.07 ± 1.87

[a] NM: not mentioned.

[b] The study involves 2–86 years of males and females. Only part of the study is shown here.

d. There was no consistency among the strength measurement protocols in the studies. Although the protocol developed by Caldwell et al. (1974) is considered the most scientific and the standard for static strength measurement, only a few studies followed it.
e. Only several studies have investigated hand torque strength at actual hand–handle interfaces, such as on hand tools like screwdrivers and wrenches, cylinders simulating handles, jar lids, electrical connectors, and small knobs.
f. There were only two studies that investigated key and oval handle, which are actually widely used in daily life and at work.
g. Most of the studies measured torque strength in standard fixed standing posture, and only few of them used free-standing posture. Only two of the studies attempted to measure the torque in both standing and sitting postures and compared them.
h. Some measured torque in clockwise, some others in counterclockwise, and still some others in both directions.
i. Only a few studies measured two-handed torque strength.

As can be seen, there are variations in certain aspects of the methodology used among the studies. Most importantly, since most of the studies did not use the standard protocol (Caldwell protocol), the accuracy of the results is questionable. Therefore, it is important that future studies follow the standard protocol for the accuracy of the results as well as comparison purposes. Dynamic torque strength studies should also be carried out. And conversion factors should be estimated between static and dynamic torque strengths for practical applications. Psychophysical torque strength studies should also be carried out to estimate psychophysical torque strength capabilities. More studies are needed for two-handed torque strength data applied to large wheels and similar handles. Postures studied are very limited; thus, varying upper extremity and body postures need to be studied simulating more realistic situations. Free-body posture seems to be more realistic that should be preferred in future studies. More studies should involve different world populations, elderly, children, and people with disabilities. Normative data are very rare. A considerable number of studies should involve generating normative data for various world populations covering wide age range and occupation groups for both genders with large number of participants.

2. The following can be driven in terms of obtained results:
   a. Gender has significant effect on hand torque strength. Caucasian females are about 60%–75% as strong as Caucasian males.
   b. Hand torque strength is about normally distributed within the population.
   c. Handle type, handle material, handle shape, handle diameter, and handle surface are all influential on hand torque strength.
   d. Effect of knurling on torque strength: results are mixed. Some studies indicate its dependency on the diameter: diameters smaller than 86 mm are ineffective.
   e. Rubber handles allow higher torque strength compared to plastic.
   f. Increasing surface contact area increases hand torque strength.
   g. Hand torque strength increases as the diameter of cylindrical handle increases and reaches maximum at 50 mm diameter; afterward, it slowly decreases.
   h. The effect of glove on the hand torque strength remains inconclusive.
   i. The higher torque strength is exerted in the standing posture than in the sitting posture. But there are also some studies finding this difference practically insignificant. In addition, exertion height has an effect on hand torque strength.
   j. Free-body posture allows higher torque strength than standard fixed-body postures.
   k. Torque strength follows a curvilinear relationship with age: in general, strength increases throughout childhood, peaks in adulthood, and then decreases with age from around 50 years.

l. In general, dominant hand is stronger than nondominant hand.
m. The effect of the direction of torque exertion remains inconclusive.
n. Manual job group is found to be significantly stronger than nonmanual group.
o. The highest torque strength values are obtained with cylindrical handle followed by circular, ellipsoid, and the lowest with key handle.
p. Torque outputs with hexagonal and circular handles are higher than triangular handles.
q. Torque capacity of subjects with carpometacarpal arthritis is lower than normal subjects.
r. 1.3 N m can be recommended as a removal torque for opening child-resistant bottle tops for Dutch population.
s. The fifth percentile torque value for females for British older adults is found 1.32 N m, whereas a torque value of 2 N m has been quoted in the literature for Dutch older adults.
t. Higher torques can be exerted on square lids compared to those that are circular of the same diameter.

Most of these results summarized are based on nonstandard strength measurement protocol; thus, some of the results are questionable. Most did not provide the statistical estimation of the sample size, so whether the sample sizes were satisfactory for drawn conclusions is unknown. The fifth percentile torque strength values still are unknown for most world populations. Children torque strength values are also unknown. Optimal circular handle diameter was estimated 50 mm. However, considering the hand length differences between genders and among individuals, the reliability of this result is questionable. As with the grip strength, one may expect that the optimal diameter of torque strength is a function of hand length. Hence, this result needs further verification.

5.4 Conclusions

It is important for ergonomists and designers to consider human capacity in design for human. In designing for hand torque strength (work or products), hand torque strength capacity of the corresponding population needs to be referred. A close examination of the torque strength studies revealed that there are important gaps in the torque strength data available to be used in work and product design. Therefore, there remains much work to do. First of all, there are only a few normative torque data available. Considering the cross-national variations, hand torque strength norms need to be developed worldwide, particularly for elderly and children and also for people with disabilities.

Besides jar opening and child-resistant removal torque strengths, the studies should continue to determine hand torque design values for other torque applications in daily life and in industrial applications. Hand torque studies with varying tool types, varying dimensions, and varying postures of body and hand need to be continued. Along with static torque strength, dynamic hand torque strength studies should also be carried out for at least to allow accurate estimations of dynamic data from static data. Furthermore, psychophysical hand torque strength studies should also be performed to determine the safe and acceptable torque levels for long-duration torque exertion tasks.

The researchers should be more serious on the use of the scientifically accepted standard torque strength procedure (i.e., Caldwell et al., 1974) and equipment as well as statistically required sample types and sizes to obtain reliable and universal results.

References

Adams, S.K., 2006. Hand grip and pinch strength. In W. Karwowski (ed.), *International Encyclopedia of Ergonomics and Human Factors*, 2nd edn., Vol. 1. CRC Press, Boca Raton, FL, pp. 365–376.

Adams, S.K., P.J. Peterson, 1986. Maximum voluntary hand grip torque for circular electrical connectors. In *Proceedings of the Human Factors Society, 30th Annual Meeting*. Human Factors Society, Santa Monica, CA, pp. 847–851.

Berns, T., 1981. The handling of consumer packaging. *Applied Ergonomics*, 12, 153–161.

Caldwell, S.L., D.B. Chaffin, F.N. Dukes-Dobos, K.H.E. Kroemer, L.L. Laubach, S.H. Snook, D.E. Wasserman, 1974. A proposed standard procedure for static muscle strength testing. *American Industrial Hygiene Association Journal*, 35(4), 201–206.

Chaffin, D., G. Andersson, B. Martin, 2006. *Occupational Biomechanics*, 4th edn. Wiley-Interscience, Hoboken, NJ.

Chaffin, D.B., 1975. Ergonomics guide for the assessment of human static strength. *American Industrial Hygiene Association Journal*, 36(7), 505–511.

Crawford, J.O., E. Wanibe, L. Nayak, 2002. The interaction between lid diameter, height and shape on wrist torque exertion in younger and older adults. *Ergonomics*, 45(13), 922–933.

Daams, B.J., 1990. Static force exertion in standardized, functional and free postures. In *Proceedings of the Human Factors Society, 34th Annual Meeting*, Santa Monica, CA, pp. 724–728.

Daams, B.J., 1994. *Human Force Exertion in User-Product Interaction.* Physical Ergonomics Series. Delft University Press, Delft, the Netherlands.

Daams, B.J., 2006. Torque data. In W. Karwowski (ed.), *International Encyclopedia of Ergonomics and Human Factors*, 2nd edn., Vol. 1. CRC Press, Boca Raton, FL, pp. 534–544.

Ekşioğlu, M., 2004. Relative optimum grip span as a function of hand anthropometry. *International Journal of Industrial Ergonomics*, 34(1), 1–12.

Ekşioğlu, M., 2006. Optimal work-rest cycles for an isometric intermittent gripping task as a function of force, posture and grip span. *Ergonomics*, 49(2), 180–201.

Ekşioğlu, M., 2011. Endurance time of grip-force as a function of grip-span and arm posture. *International Journal of Industrial Ergonomics*, 41(5), 401–409.

Ekşioğlu, M., E. Baştürk, 2013. An estimation of isometric hand torque strength of adult male population of turkey and effects of various factors. In *Proceedings of International IIE Conference & YAEM 2013*, İstanbul, Turkey, pp. 111–112.

Ekşioğlu, M., K. Kızılaslan, 2008. Steering-wheel grip force characteristics of drivers as a function of gender, speed, and road condition. *International Journal of Industrial Ergonomics*, 38, 354–361.

Ekşioğlu, M., Z. Recep, 2013. Hand torque strength of female population of turkey and the effects of various factors. In P. Arezes, J.S. Baptista, M.P. Barroso, P. Carneiro, P. Cordeiro, N. Costa, R.B. Melo, A.S. Miguel, G. Perestrelo (eds.), *Occupational Safety and Hygiene.* CRC Press, Boca Raton, FL, pp. 37–41.

Gallagher, S., J.S. Moore, T.J. Stobbe, 1998. *Physical Strength Assessment in Ergonomics.* American Industrial Hygiene Association, Fairfax, VA.

Imrhan, S.N., C. Loo, 1986. Torque capabilities of the elderly in opening screw top containers. In *Proceedings of the Human Factors Society, 30th Annual Meeting*, Dayton, OH, Vol. 30(12), pp. 1167–1171.

Imrhan, S.N., G.D. Jenkins, 1999. Flexion-extension hand torque strengths: Applications in maintenance tasks. *International Journal of Industrial Ergonomics*, 23, 359–371.

Kim, C.H., T.K. Kim, 2000. Maximum torque exertion capabilities of Korean at varying body postures with common hand tools. In *Proceedings of the Human Factors and Ergonomics Society Annual Meeting*, San Diego, CA, Vol. 3, pp. 157–160.

Kong, Y.K., B.D. Lowe, S.J. Lee, E.F. Krieg, 2007. Evaluation of handle design characteristics in a maximum screwdriving torque task. *Ergonomics*, 50(9), 1404–1418.

Kroemer, K., 2006. Static and dynamic strength. In W. Karwowski (ed.), *International Encyclopedia of Ergonomics and Human Factors*, 2nd edn., Vol. 1. CRC Press, Boca Raton, FL, pp. 511–512.

Kroemer, K.H.E., 1970. Human strength: Terminology, measurement and interpretation of data. *Human Factors*, 12(3), 297–313.

Kroemer, K.H.E., H.J. Kromer, K.E. Kroemer-Elbert, 1997. *Engineering Physiology*, 3rd edn. Van Nostrand Reinhold, p. 107.

Leigh, J.P., 2011. Economic burden of occupational injury and illness in the United States. *The Milibank Quarterly*, 89(4), 728–772.

Miller, M.C., M. Nair, M.E. Baratz, 2005. A device for assessment of hand and wrist coronal plane strength. *Journal of Biomechanical Engineering*, 127, 998–1000.

Mital, A., 1986. Effect of body posture and common hand tools on peak torque exertion capabilities. *Applied Ergonomics*, 17(2), 87–96.

Mital, A., S. Kumar, 1998. Human muscle strength definitions, measurement and usage: Part I—Guidelines for the practitioner. *International Journal of Industrial Ergonomics*, 22, 101–121.

Mital, A., N. Sanghavi, 1986. Comparison of maximum volitional torque exertion capabilities of males and females using common hand tools. *Human Factors*, 28(3), 283–294.

Nagashima, K., S. Konz, 1986. Jar lids: Effect of diameter, gripping materials and knurling. In *Proceeding of the Human Factors Society, 30th Annual Meeting*. Human Factors Society, Santa Monica, CA, pp. 672–674.

Nayak, U., J. Queiroga, 2004. Pinch grip, power grip and wrist twisting strengths of community-dwelling, healthy older adults. *Gerontechnology*, 3(2), 77–88.

Norris, B., J.R. Wilson, 1997. *Designing Safety into Products*. Product Safety and Testing Group, Institute for Occupational Ergonomics, London, U.K.

Peebles, L., B. Norris, 2003. Filling 'gaps' in strength data for design. *Applied Ergonomics*, 34, 73–88.

Pheasant, S., D. O'Neill, 1975. Performance in gripping and turning—A study in hand/handle effectiveness. *Applied Ergonomics*, 6(4), 205–208.

Replogle, J.O., 1983. Hand torque strength with cylindrical handles. In *Proceedings of the Human Factors Society, 27th Annual Meeting*. Human Factors Society, Santa Monica, CA, pp. 412–416.

Rohles, F.H., K.L. Moldrup, J.E. Laviana, 1983. *Opening Jars: An Anthropometric Study of the Wrist Twisting Strength of Children and the Elderly*. Institute for Environmental Research, Kansas State University, Manhattan, KS.

Rowson, J.A., A. Yoxall, 2011. Hold, grasp, clutch or grab: Consumer grip choices during food container opening. *Applied Ergonomics*, 42, 627–633.

Schoorlemmer, W., H. Kanis, 1992. Operation of controls on everyday products. In *Proceedings of the Human Factors, 36th Annual Meeting*. Human Factors and Ergonomics Society, Santa Monica, CA, pp. 509–13.

Seo, N.J., T.J. Armstrong, J.A. Ashton-Miller, D.B. Chaffin, 2007. The effect of torque direction and cylindrical handle diameter on the coupling between the hand and a cylindrical handle. *Journal of Biomechanics*, 40, 3236–3243.

Seo, N.J., T.J. Armstrong, D.B. Chaffin, J.A. Ashton-Miller, 2008. The effect of handle friction and inward or outward torque on maximum axial push force. *Human Factors*, 50, 227–236.

Smith, J.L., 2006. Static muscle strength. In W. Karwowski (ed.), *International Encyclopedia of Ergonomics and Human Factors*, 2nd edn., Vol. 1. CRC Press, Boca Raton, FL, pp. 513–514.

Snook, S.H., 1985. Psychophysical acceptability as a constraint in manual working capacity. *Ergonomics*, 28, 331–335.

Steenbekkers, L.P.A., 1993. *Child Development, Design Implications and Accident Prevention*. Physical Ergonomics Series. Delft University Press, Delft, the Netherlands.

Swain, A.D., G.C. Shelton, L.V. Rigby, 1970. Maximum torque for small knobs operated with and without gloves. *Ergonomics*, 3(2), 201–208.

Voorbij, A.I.M., L.P.A. Steenbekkers, 2002. The twisting force of aged consumers when opening a jar. *Applied Ergonomics*, 33, 105–109.

Wieszczyk, S.M., R.W. Marklin, H.J. Sánchez, 2009. Height of industrial hand wheel valves affects torque exertion. *Human Factors*, 51, 487–496.

Yoxall, A., R. Janson, S.R. Bradbury, J. Langley, J. Wearn, S. Hayes, 2006. Openability: Producing design limits for consumer packaging. *Packaging Technology and Science*, 19, 219–225.

6

Optimization of Traffic Flow on Kuwait's Roads and Highways

CHIBLI JOUMAA, NORIAH AL-MAS, SUAD AL-SUWAITI, NOOR ASHOUR, AND SHAIMA GOODARZI

Contents

6.1 Introduction

As a resident of the state of Kuwait, my main focus of this study is the increasing traffic problems on the Kuwait roads. Traffic is taking its turn to the worst, and if not handled, it will continue to rise. The aim of this chapter is to study and optimize the traffic flow and reduce congestions as much as possible in Kuwait.

The population of Kuwait has been increasing. According to the Kuwait government [1,2], the number of residents in Kuwait went from 321,621 in 1961 to 1,697,301 in 1985, to 2,213,403 in 2005, and to 3,328,136 in June 2008. Naturally, this large increase throughout the years led to the increase in the number of automobiles circulating in the country. As a matter of fact, according to a governmental study [2], Kuwait witnesses around 6%–9% annual increase in the number of vehicles. Since the area of Kuwait is rather small, the increase of cars resulted in traffic congestion. The study also showed that the population growth would continue to increase, and it is expected to have more than 8,000,000 residents by 2020, meaning that the traffic problems will only get worse. That is why optimizing the traffic in Kuwait has become a necessity.

In Section 6.2, literature review, including survey development and algorithms for network optimization and transportation models that are to be applied to Kuwait's network are presented. These methods have been derived from several operations and research books, as well as scientific papers, and have therefore been applied.

Section 6.3 consists of a theoretical approach used to gather all the information required to analyze the existing flow of traffic. Gathering the information was the longest process. The data gathering process took about 2 months to develop, distribute, and collect the results. Surveys were developed to accommodate the public, in the sense of making it as easy as possible for people to follow. The idea of online surveys helped speed up the process in which surveys were distributed through the social media; therefore, the responses were faster. To approach the issue, studying the existing road divisions of Kuwait was required.

Dividing Kuwait into different zones was done using both the existing road network division of Kuwait and the results of the surveys. An observation of the main roads, highways, areas, and people's behavior was made in order to study the flow of automobiles. Finally, a network was created for Kuwait, which linked all of the zones together using arcs and nodes representing roads and areas, respectively.

In Section 6.4, results and discussions, including the analysis of the surveys along with the statistical analysis of traffic is presented for every time period of the day. Each time period has a dedicated map that shows the flow of traffic and the zones that are busy. Then simulation through the minimal spanning tree would indicate the shortest route from one zone to another, linking all zones in Kuwait. Finally, scenarios were conducted and improvements made for special cases. It was important to conduct scenarios to show the natural flow, where improvements were made even when misfortunes would occur.

The chapter ends with the conclusion. Appendix 6.A includes paper-based and online surveys, which is followed by references.

6.2 Literature Review

Literature review calls attention to critical points of knowledge of all information and sources being used for the study made in certain fields and also evaluates and designates the path of the chapter. Literature review is a summary and outline of a specific area of research, indicates the reason and aims for perusing this research made, and aids the reader to clearly understand what the chapter is about.

6.2.1 Survey Development

In order to study the traffic flow and analyze and optimize it, information about the traffic is required. Ref. [3] offers sample survey questions, answers, and tips to those who incorporate surveys into projects or services. One of the first important aspects when developing surveys is to ultimately satisfy customers; sample questions are offered for inspiration, one of which is to define gender and include age ranges. The incorporation emphasizes on the fact that answers should be provided in a form of bullet points, in which the answers are clear when chosen.

By definition, a survey studies a sample of individuals from a given population and draws conclusions about the population based on the sample. Therefore, for accurate and reliable results, the sample to be studied must be defined correctly. The number of samples should represent well the whole population.

6.2.2 *Algorithms*

Linear programming [4–10] is used for decision-making purposes. All linear programming models include three components:

1. Decision variables that we want to determine, which is the first step in developing a model.
2. The objective function (goal) that will be optimized by either maximizing or minimizing:

$$\text{Maximize/Minimize } c_1x_1 + c_2x_2 + c_3x_3, \cdots, c_nx_n.$$

3. The constraints [4] that are used to limit the values, and the solution must satisfy those constraints:

$$a_{11}x_1 + a_{12}x_2 \le b_1,$$

$$a_{21}x_1 + a_{22}x_2 \le b_2,$$

$$a_{31}x_1 + a_{32}x_2 \le b_3.$$

Any values of the defined variables that satisfy all the constraints give a feasible solution. Otherwise, the solution is infeasible. However, the goal of linear programming is to find the optimum, which is the best feasible solution that either maximizes or minimizes the objective function [4].

6.2.2.1 Network Model In this section, we introduce network optimization algorithms. Many algorithms were developed over the years; we will focus on the following:

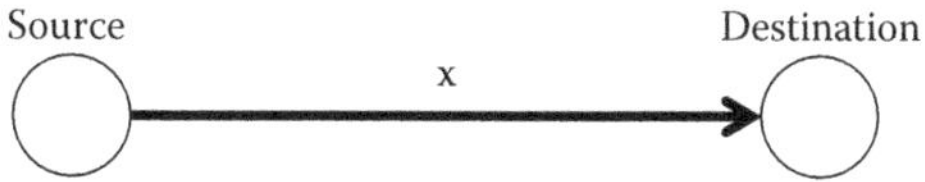

Figure 6.1 Nodes and arcs.

1. Minimal spanning tree
2. Shortest route
   a. Dijkstra's algorithm
   b. Floyd's algorithm

A network consists of a set of nodes and arcs. The notation for describing a network model is (N, A), where N is the set of nodes and A is the set of arcs. Nodes are sources and destinations, and the arcs represent the flow between the nodes (Figure 6.1). A network is considered connected if there are two nodes linked by a path. The flow on the arcs in our case will be the automobile traffic flow in highways [4].

6.2.2.1.1 Minimal Spanning Tree Algorithm The minimal spanning tree algorithm links the nodes of a network using the shortest link possible [5,7]. The advantage of the minimal spanning tree solution is that it provides the most economical design of the road system.

Nodes may represent areas, intersections, and bus stops, while the links can represent the capacity of the route, cost, or distance. This algorithm is used in both transportation and communication infrastructures to reach the optimal network [6].

Algorithm:

Step 1: Select the shortest link between any two given nodes in the network.

Step 2: Select the shortest potential link between a node that has already been touched by a link and a node that has not yet been connected.

Step 3: Repeat step 2 until every node is touched by a link and until the final destination is reached.

6.2.2.1.2 Shortest Route Algorithm In a transportation network, the shortest route algorithm determines the shortest route between two nodes, a source and a destination.

Two algorithms are used for solving such networks [4].

6.2.2.1.2.1 Dijkstra's Algorithm This algorithm is used to determine the shortest routes between the source node and every other node in the network. Dijkstra's algorithm can be applied to improve and optimize the logistics distribution system. In result, the efficiency of transport vehicles will be improved, and man and material resources will be saved.

Algorithm:

Step 1: Label the first node (source) with a zero label.
Step 2: Add the cost on the arc, between node zero and the next destination, with node zero and place the result in the destination node.
Step 3: Repeat step 2, but this time, the source will be the destination with the shortest route reached by the last step.
Step 4: Repeat previous steps until you reached the final desired destination.

6.2.2.1.2.2 Floyd's Algorithm Floyd's algorithm is used for determining the shortest route between any two nodes in the network. It is a more general approach compared with Dijkstra's algorithm because it just requires two nodes from one network. It is used for freight trains that use railroad network to transport goods from an origin to a destination [9].

Algorithm:

Step 1: Start by drawing two tables containing rows and columns depending on the number of nodes, one for the distance (D_0) and one for the nodes (S_0).
Step 2: Highlight the first column and row of table D_1. In the unhighlighted cells, search for the infinity sign if available; if not, search for the highest number.
Step 3: The cell that has the infinity sign should be changed to the sum of its highlighted coordinates. The cell that has been changed in D_1 table should be changed to the iteration K in table S_1.
Step 4: Repeat steps 2 and 3, but instead of highlighting the first row and column, highlight the second and then third until you reach the last column or row.

After editing the tables, the end result would be the shortest distance between each source and destination.

6.3 Methodology

This section represents the approach leading to the results and improvements of Kuwait's roads. The first step is the data collection process. Provided in Section 6.3.1 is how surveys were developed and data gathered through online and paper-based surveys. As mentioned earlier, Kuwait's population is above three million. A sample population should be defined. The minimum number of collected surveys should be determined in order to ensure the representativeness of the sample with regard to Kuwait's population. Afterward, the data regarding Kuwait's road lengths, capacities, and speed information will be computed. After the calculations regarding Kuwait's roads are made, Kuwait is divided into zones according to the highways and their intersections. Finally, Kuwait's network is built.

6.3.1 Data Collection

The first step taken was the data collection process. Surveys were developed, distributed, and collected. Before completing the data collection process, the minimum sample size required to have accurate results needs to be calculated. Several methods exist to define the sample size to be used. In this study, the *sample size infinite population* formula was used from [11], where the population should exceed 50,000.

The minimum sample size required is calculated using the following formula:

$$n = \frac{Z^2 \times (p) \times (1-p)}{C^2}, \tag{6.1}$$

where

n is the sample size
Z is the Z-value (1.645 = 90% confidence level)
p is the percentage of population expected to respond (generally 0.5)
C is the confidence interval, expressed as decimal (0.6, it can range between 4% and 6%)

The minimum sample size expected is 187.9 = 188 answered surveys.

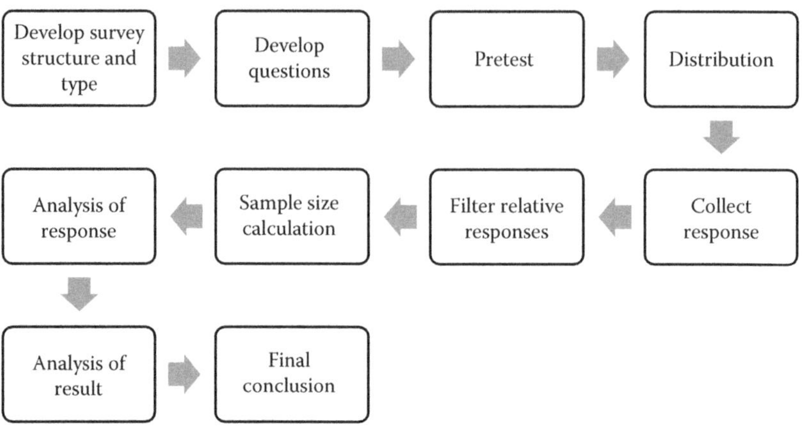

Figure 6.2 Constructing a survey.

After computing the minimum number of surveys needed, two types of surveys were developed: paper-based and online surveys. A sample of these surveys can be found in Appendix 6.A.

The information needed for the study was defined, and questions were tailored accordingly in the surveys. Using the surveys, we can specify the routes, sources, and destinations of the people at given times and hence track their movements.

One hundred and nine paper-based surveys and one hundred and thirty-one online surveys were answered. This gives a total of 240 surveys, which is above the minimum calculated sample size.

The steps followed are shown in Figure 6.2.

6.3.1.1 Developing Questions Certain questions were developed for the purpose of tracking the congestions of roads and zones. Those questions include the timing of departure from sources and arrival at destinations. In addition, people were asked about the routes chosen by them in order to determine the trend.

The whole process took approximately 2 months, which included developing, distributing, and gathering. At first, only paper-based surveys were used and distributed to various people. However, it was clear that a more random survey and a wider population were needed; therefore, the idea of creating an online survey was introduced. Online surveys were simpler because the main goal was for the people to complete it as fast as possible. As for the distribution of the surveys, the online version was spread through social networks. The online survey sample is shown in Appendix 6.A.2.

After all the data were collected, they were put in an Excel sheet, which would make the organization and analysis of them easier.

6.3.1.2 Paper-Based Surveys To locate the traffic in Kuwait roads and to know what areas are frequently visited, a survey was distributed to the public. The goal of the survey was to know the source, the destination, and the links between them, which are roads. Knowing the time of leaving from and arriving at a certain area made it easier to calculate the time spent on the road. The paper-based survey was basically a combination of three different surveys, one addressing workdays, one addressing after-work movements, and one addressing the weekend. A sample survey is provided in Appendix 6.A.1.

Two hundred and thirty paper-based surveys were collected, but only one hundred and nine surveys were counted in the analysis due to the fact that some of them were incomplete. Therefore, the response was 47.39%.

6.3.1.3 Online-Based Surveys After the paper-based surveys were carried out, a broader sample of people was needed. Therefore, an online survey was created using Survey Monkey, 2012 and was randomly distributed to the general public in Kuwait using social networks. The purpose of this survey was to distinguish where people during specific time periods were.

Just like the paper-based surveys, the online survey differentiated between workdays and weekends. For each section, the time periods were specified, and people stated which area they were in during that time, which helped identify the areas that were the most crowded during the time periods. The results also helped in dividing Kuwait into zones and pointing out the congested zones.

6.3.2 Road Information

Capacity analysis involves quantitative evaluation of the capacity of a road section. It uses a set of procedures to determine the maximum flow of traffic that a section would carry [12].

Possible capacity is defined as the maximum number of vehicles that can pass a point in 1 h under prevailing roadway and traffic condition. Practical capacity, on the other hand, is the maximum number of vehicles that can pass a point without restrictions for average driver's to pass other vehicles.

The main goal of this section is to find out the number of cars a road can handle. Therefore, information regarding the roads was gathered, and numerous calculations were conducted. The Ministry of Interior provided the information needed for conducting the capacity calculations, which involve the length of the roads, the maximum speed, and the minimum speed. The average speed was computed along with the time for one car to cross the road. Equation 6.2 illustrates the time for one car to cross the road.

T is the time for one car to cross the road (h):

$$T = \frac{L}{A}, \tag{6.2}$$

where

A is the average speed for each lane (km/h)
L is the length of the road (km)

The next step is to find the capacity of the lane to ultimately compute the actual number of cars on the road. The average length of a car was found through measuring several cars to eventually find out how many cars fit a lane.

To find the capacity per lane, the average length of cars (C), as well as the space provided between the cars (S), is needed.

Average length of a car = 0.00443 km (three cars of 0.0043, 0.0043, and 0.0047 km were measured)

$$\text{Capacity per lane} = \frac{R}{(C+S)}, \tag{6.3}$$

where

R is the road length (km)
C is the average car length (km)
S is the space between each car (0.001 km)

Table 6.1 shows information on the road, whereas Table 6.2 shows information on the capacity of the roads. The time taken for

Table 6.1 Road Information

STREET NAME	LENGTH (km)	MAXIMUM SPEED (km/h)	MINIMUM SPEED (km/h)	AVERAGE SPEED (km/h)	TIME FOR ONE CAR TO CROSS THE ROAD (T in h)
1st Ring Road	4.7	100	40	70	0.067
2nd Ring Road	8	80	40	60	0.133
3rd Ring Road	8.3	80	40	60	0.138
4th Ring Road	15	80	40	60	0.250
5th Ring Road	30	120 (bridge between Rumaithiya and Salmiya = 80)	40	80	0.375
6th Ring Road	58	120	40	80	0.725
7th Ring Road	32	120	40	80	0.400
Fahaheel Highway (starting from south until 4th Ring Road)	50	120	60	90	0.556
Fahaheel (after 4th Ring Road)	6.4	80	60	70	0.091
King Fahad (starting from Saudi Arabia until 5th Ring Road)	96.5	120	60	90	1.072
King Fahad (after 5th Ring Road until Kuwait City)	7	100	60	80	0.088
King Faisal (from Airport to Qortuba)	2.5	120	60	90	0.028
King Faisal (from Qortuba until 1st Ring Road)	1.5	100	60	80	0.019
Al-Ghazally Street	10	100	60	80	0.125
Damascus Street	7	80	40	60	0.117

Table 6.2 Capacity Information

STREET NAME	CAPACITY PER LANE (CARS/LANE)	NUMBER OF LANES
1st Ring Road	866	2
2nd Ring Road	1,474	3
3rd Ring Road	1,529	3
4th Ring Road	2,763	3
5th Ring Road	5,525	3
6th Ring Road	10,682	3
7th Ring Road	5,894	4
Fahaheel Highway (starting from south until 4th Ring Road)	9,209	3
Fahaheel (after 4th Ring Road)	1,179	4
King Fahad (starting from Saudi Arabia until 5th Ring Road)	17,772	3
King Fahad (after 5th Ring Road until Kuwait City)	1,290	4
King Faisal (from Airport to Qortuba)	461	3
King Faisal (from Qortuba until 1st Ring Road)	277	3
Al-Ghazally Street	1,842	2
Damascus Street	1,290	3

one car (T) to cross the road was shown in Table 6.1. The next step is to determine how many cars are able to cross the road in a specific period of time. The time was determined through the specified time slots that were used during the data collection process.

The calculation up until now has been directed toward an ideal case; since we are living in the real-world, estimations were added to the formula. Therefore, percentage was added to the road capacity to ensure that despite the actual cars that passed through, there were a certain number of cars that made it to their destination [9].

Our estimation was that 40% of the cars would pass the road and reach their desired destinations:

$$N = (P + (40\% \times \text{capacity per lane})) \times L, \tag{6.4}$$

where

N is the actual number of cars per lane
L is the number of lanes
P are the cars in a specified period = (h/period)/T

Table 6.3 shows the number of cars in each road.

Table 6.3 Capacity of Each Road per Period

STREET NAME	6–8 A.M.	8 A.M.–12 P.M.	12–4 P.M.	4–6 P.M.	6–11 P.M.
1st Ring Road	753	812	812	753	842
2nd Ring Road	1,814	1,859	1,859	1,814	1,882
3rd Ring Road	1,878	1,922	1,922	1,878	1,943
4th Ring Road	3,340	3,364	3,364	3,340	3,376
5th Ring Road	6,646	6,662	6,662	6,646	6,670
6th Ring Road	12,827	12,835	12,835	12,827	12,839
7th Ring Road	9,450	9,470	9,470	9,450	9,480
Fahaheel Highway (starting from south until 4th Ring Road)	11,062	11,072	11,072	11,062	11,078
Fahaheel (after 4th Ring Road)	1,974	2,062	2,062	1,974	2,106
King Fahad (starting from Saudi Arabia until 5th Ring Road)	21,332	21,338	21,338	21,332	21,340
King Fahad (after 5th Ring Road until Kuwait City)	2,155	2,246	2,246	2,155	2,291
King Faisal (from Airport to Qortuba)	767	982	982	767	1,089
King Faisal (from Qortuba until 1st Ring Road)	648	964	964	648	1,122
Al-Ghazally Street	1,506	1,538	1,538	1,506	1,554
Damascus Street	1,599	1,651	1,651	1,599	1,676

6.3.3 Dividing Kuwait's Map

Kuwait's roads are designed to cover the whole country. According to the map of Kuwait, there are roads that are well defined; they include horizontal ring roads, and roads and highways that intersect vertically through the ring roads. Seven horizontal ring roads are named sequentially starting from 1 to 7.

- *The 1st Ring Road*—which starts from Sharq and ends at Qiblah.
- *The 2nd Ring Road*—which starts from Bnaid Al-Gar and ends at Shuwaikh.
- *The 3rd Ring Road*—which starts from Al-Daiya and ends at Kaifan.
- *The 4th Ring Road*—which starts from Salmiya and ends at Shuwaikh Industrial.

- *The 5th Ring Road*—which starts from Salmiya and ends at Doha.
- *The 6th Ring Road*—which starts from Mishref and ends at Jahra.
- *The 7th Ring Road*—which starts from Mubarak Al-Kabeer and extends to the west of Kuwait.

The vertical roads that intersect the ring roads are as follows:

- *Gulf Road*—which starts from Salmiya and ends at Shuwaikh.
- *Fahaheel/Istiqlal* (30)—which starts from Kuwait City and ends at Chalets District.
- *King Fahad/Maghreb* (40)—which starts from Kuwait City and ends at Kuwait/Saudi Arabia Borders.
- *King Faisal/Riyadh* (50)—which starts from Kuwait City and ends at Al-Ahmadi.
- *Airport Road* (55)—which starts from Shuwaikh Industrial and ends at the Kuwait International Airport.
- *Ghazalli Road* (6)—which starts from Shuwakh Residential and ends at the Airport.
- *Al-Jahra Road* (80)—which starts from Kuwait City and ends at Al-Jahra.

Based on the ring roads and intersections, the map was divided into various zones. On the map and the survey results, the first identifications were the main roads and areas of attraction.

Kuwait City was clearly the biggest issue, so it was taken as a separate zone. Moreover, The Avenues (the largest mall in Kuwait, which is a very popular attraction), Jabriya (a residential area where there are some famous schools and hospitals), and Kuwait University were also some of the congested areas on weekends and weekdays. Finally, 48 zones were created.

Figure 6.3 describes the chosen ring roads and highways.

As shown in Figure 6.4 extracted from Ref. [13], the marked destinations are identified due to the intersections made by main roads through each ring road. These intersections connect main roads and divide areas beneath the ring roads; therefore, the divided areas were chosen as zones. The red borders in Figure 6.4 shows how Al-Adailiya is selected as a zone, as it lies between the 3rd

	Ghazali (6)	Airport road (55)	King Faisal	Damascus	King Fahad (40)	Faheel (30)	Gulf road
First ring road							
Second ring road							
Third ring road							
Fourth ring road							
Fifth ring road							
Sixth ring road							
Seventh ring road							

Figure 6.3 Representation of Kuwait's Roads.

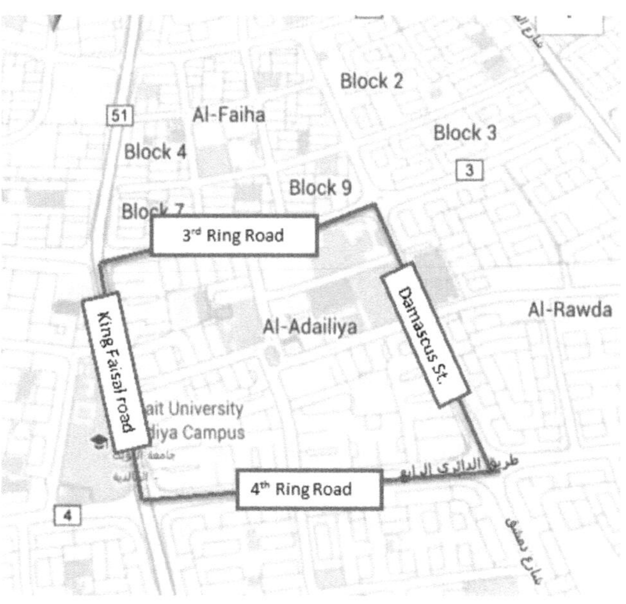

Figure 6.4 Method used to divide zones.

and 4th Ring Roads and is intersected by King Faisal Road and Damascus Street as indicated.

The zones are selected with respect to the results of the surveys concerning the attraction areas and roads that are often used. The zones are summarized in Table 6.4.

Table 6.4 Zones

ZONES	AREAS
1	Al-Zahra
2	Abdullah Al-Salem
3	Abraq Khaitan
4	Adan + Qusoor + Qurain + Mubarak Al-Kabeer + Sabah Al-Salem + Egaila + Al-Rigga + Sabahiya
5	Airport District + Al-Dhajeej + Abdullah Al-Mubarak + Jleeb Al-Shuyoukh
6	Al-Abdli
7	Al-Adailiya
8	Al-Ahmadi
9	Al-Bidea
10	Al-Daiya + Qadsiya
11	Al-Farwaniya
12	Al-Jahra + Sa'ad Al-Abdullah
13	Al-Khafji + Al-Khiran + Bnaider + Jlai'a + Al-Zoor
14	Al-Mansouriyah + Dasma
15	Al-Messila + Abu Al-Hassani + Mahboula + Fintas + Abu Halifa + Al-Mangaf + Fahaheel
16	Al-Rabiya + Al-Rahab + Ishbilya + Ardiya
17	Al-Rai
18	Al-Rawda
19	Kabd
20	Al-Sha'ab
21	Al-Shamiya
22	Al-Wafra
23	Andalus + Riggae + Nahda
24	Bayan + Mishref
25	Bnaid Al-Gar
26	Dahar
27	Faiha
28	Firdous + Sabah Al-Nasser
29	Hawalli
30	Jabriya
31	Kaifan
32	Khaldiya
33	Kuwait City (Murgab + Sharq + Kuwait City)
34	Mina Abdullah
35	Nuzha
36	Qayrawan
37	Qortuba

(*Continued*)

Table 6.4 (*Continued*) Zones

ZONES	AREAS
38	Qurnata
39	Rumaithiya + Salwa
40	Salmiya
41	Shuwaikh A (Residential)
42	Shuwaikh B (Industrial)
43	Shuwaikh C (Health Region + Kuwait University)
44	South Surra
45	Subhan
46	Sulaibiya
47	Surra
48	Yarmouk

6.3.4 Building Kuwait's Network

A network model of Kuwait is built in order to simulate the current road situation. When building a network, nodes and arcs are needed. The nodes are represented by zones, which are either sources or destinations. The arcs represent the routes that link zones together.

To construct the network, each zone has been taken and directly linked to a neighboring zone. The route has been recorded, and the distance and the capacity have been calculated. The data are then used to simulate several optimization algorithms. Figure 6.5 shows a sample of Kuwait's network.

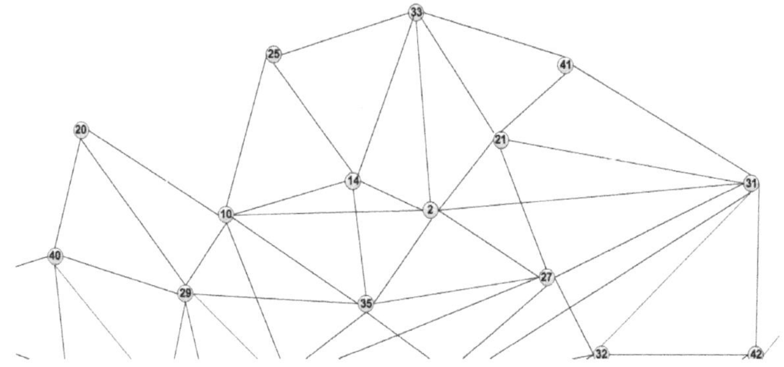

Figure 6.5 Kuwait's network sample.

6.4 Results and Discussion

6.4.1 Analyzing Results of Surveys

6.4.1.1 General Survey Analysis The sample population for the surveys is described in Table 6.5. Males answered 46.7% of the surveys, and females answered 53.3%. Note that surveys were distributed randomly between males and females, of different age groups and different occupations. This variety will help get more accurate and reliable results. The majority of the participants' age groups (95%) were 18 and above; therefore, the data gathered were for people driving and owning cars.

The Ministry of Interior generously provided us with the day's peak times, and after the surveys were analyzed and the behavior of people during the day studied, the day was divided into five periods:

1. 6–8 a.m.: Mostly students and employees leave during this time.
2. 8 a.m.–12 p.m.: The range of time at which people arrive to work depending on their shifts.
3. 12–4 p.m.: The range of time at which people arrive or leave work depending on their shifts.
4. 4–6 p.m.: Most people leave work or their house to run errands.
5. 6–11 p.m.: People are busy with their daily activities.

6.4.1.2 Statistical Analysis of Traffic Grouping data were required for the analysis of results from the surveys. Areas in Kuwait should be ranked according to the number of people in that area and in

Table 6.5 Survey Participants

Number of people participated			
Male			Female
112			128
Age group of participants			
Below 18	18–25	26–45	Above 45
3	103	91	34
Occupation of participants			
Students	Employees	Retired	Other
77	149	3	9

neighboring main routes. One approach was to group the data into two elements—congested and not congested. This method is inaccurate since it lacks detail on the levels of congestion.

For each time period in both weekdays and weekends, results from the surveys were grouped using cluster analysis. Specifically, the cluster analysis method used was hierarchical cluster analysis in the IBM SPSS Statistics software to group data as sources, destinations, and their routes. This method has been conducted to set data in clusters that will aid in visualizing and analyzing the level of congestion in each zone and road.

Before beginning the cluster analysis, the number of clusters should be determined. Starting with two clusters—congested and not congested—lots of detail has been eliminated. Going to three clusters, this number is compared with having a number of four clusters.

Figure 6.6 shows the result during the weekdays from 6 to 8 a.m. The map on the left in Figure 6.6 represents three clusters, and the second map represents four clusters. Working with three clusters, zones that are congested are shown to have low congestion, such as zone 33 (Kuwait City). With four clusters, zone 33 has shifted from

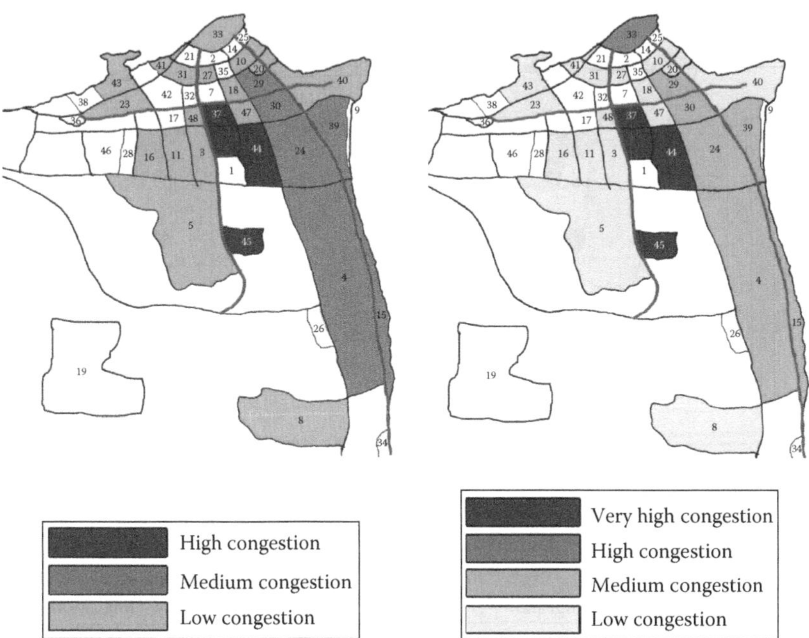

Figure 6.6 Comparison between three and four clusters.

having low congestion to high congestion. Therefore, more detail is being added, and crowded zones are appearing with four clusters.

In addition, a mathematical approach is used. In order to divide the data into clusters, the optimal number of clusters k was determined using Equation 6.5 [14]:

$$K = \sqrt{\frac{n}{2}}, \tag{6.5}$$

where n is the number of data points.

When Kuwait was divided into 48 zones, the number of clusters k was 4.8. From the choice of four or five groups, four clusters were been chosen. As seen in Figure 6.7, with five groups, little detail was added. Zone (30) Jabriya changed colors; however, it is still considered medium congested.

With four groups chosen, Figure 6.8 shows the results of the weekdays with four clusters.

The three most congested routes in this time period are Fahaheel Motorway (30) with 20% congestion level compared with all road

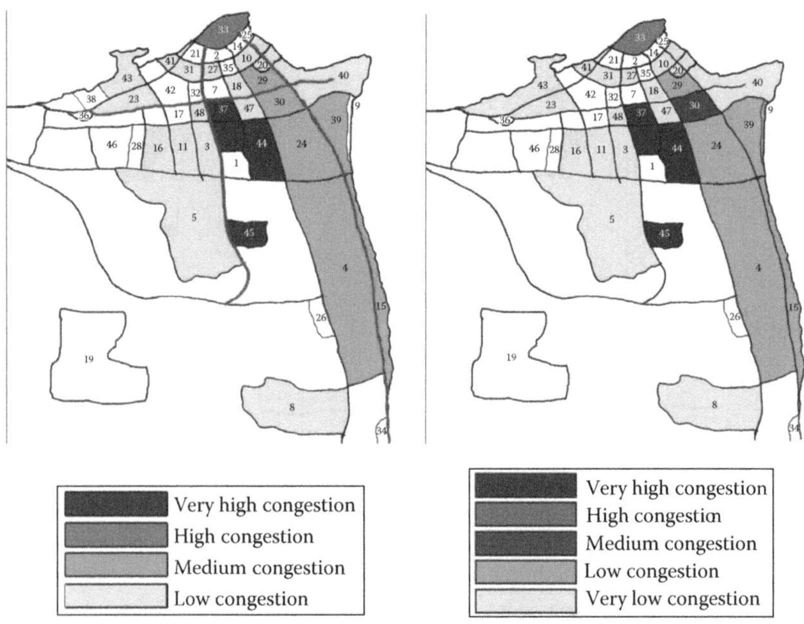

Figure 6.7 Comparison between four and five clusters.

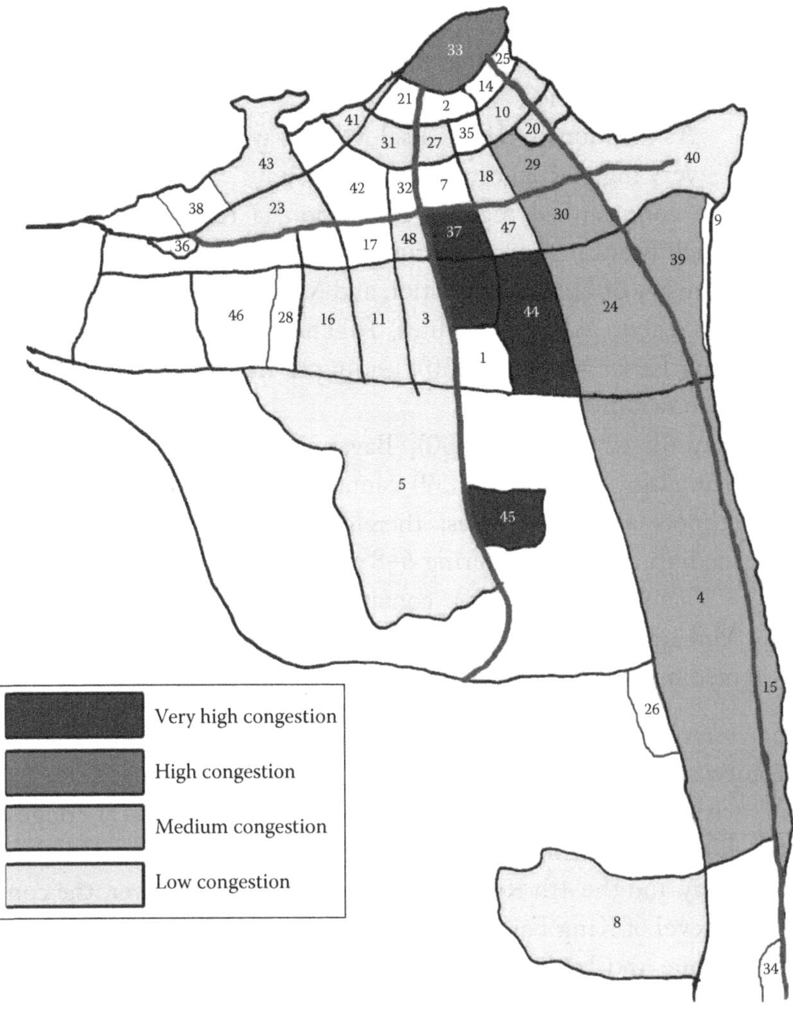

Figure 6.8 Weekdays, 6–8 a.m.

activities, the 4th Ring Road with 13.61% congestion, and finally King Faisal Motorway—Riyadh (50)—with 11.52% of the crowding level:

1. The zones ranked high in congestion are as follows:
   a. *Residential areas*: South Surra (44) and Qortuba (37). During this time period, occupants leave their houses to go to work or to drop their children to school in the same or neighboring zones.

   b. *Major organizations*: Subhan (45). This location has major companies, including oil companies and, mostly, Kuwait-based factories.
      As mentioned, King Faisal Motorway (50) passing by the very highly congested zones is busy.
2. The zone ranked high in congestion is Kuwait City (33), which is destination point for multiple ministries such as the Ministry of Higher Education and Ministry of Social Affairs and Labor. Also, the routes, Fahaheel Motorway (30) and King Faisal Motorway (50), leading to this area are crowded as shown in Figure 6.8.
3. Hawalli (29), Jabriya (30), Bayan and Mishref (24), and Rumaithiya and Salwa (39) contain both public and private schools and universities; therefore, they are ranked with medium congestion during 6–8 a.m.
   Zones (4) and (15) containing residential areas like Mubarak Al-Kabeer, Abu Halifa, and Al-Qusoor are also medium congested.

The congested routes represented in Figure 6.9 are again Fahaheel Motorway (30) with 15.38% congestion, King Faisal Motorway (50) with 26.92%, and the 4th Ring Road with 11.54% congestion. During this time period, the number of people on Fahaheel Motorway and the 4th Ring Road has decreased; however, the congestion level of King Faisal Motorway passing by a very highly congested zone and leading to a highly congested zone has increased by 15.4%:

1. The zone ranked very high in congestion is South Surra (44); besides being a residential area, this area also includes ministries such as the Ministry of Electricity and Water and the Public Authority for Civil Information. Clients come to this area for their paperwork.
2. Kuwait City (33) is again ranked high in congestion. Routes (30) and (50) leading to this area are also busy.

Figure 6.10 shows that the most congested routes during weekdays from 12 till 4 p.m. are Fahaheel Motorway (30) with the highest percentage of 15.72, King Faisal Motorway (50) with

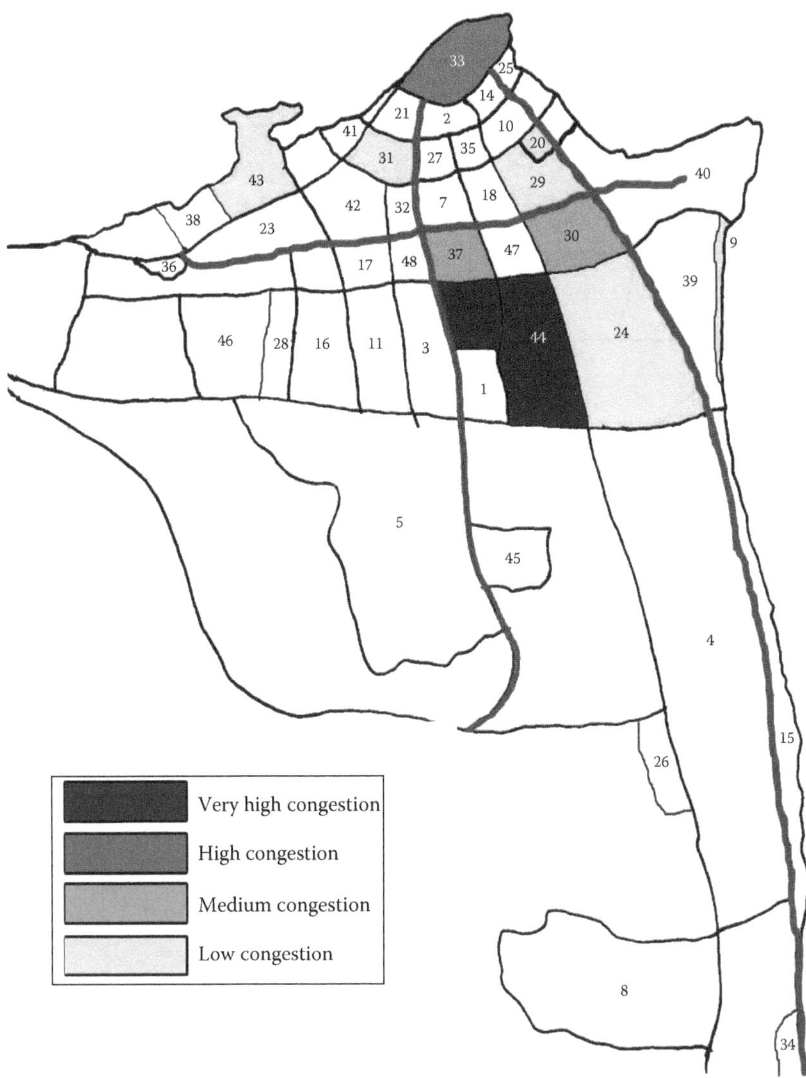

Figure 6.9 Weekdays, 8 a.m.–12 p.m.

13.84%, and King Fahad Motorway (40) with 10.69% of the total roads' activity:

1. The zone ranked very high in congestion is only Qortuba (37). During this time period, occupants leave their work and start heading home. Others also may start leaving for work for their second shift.

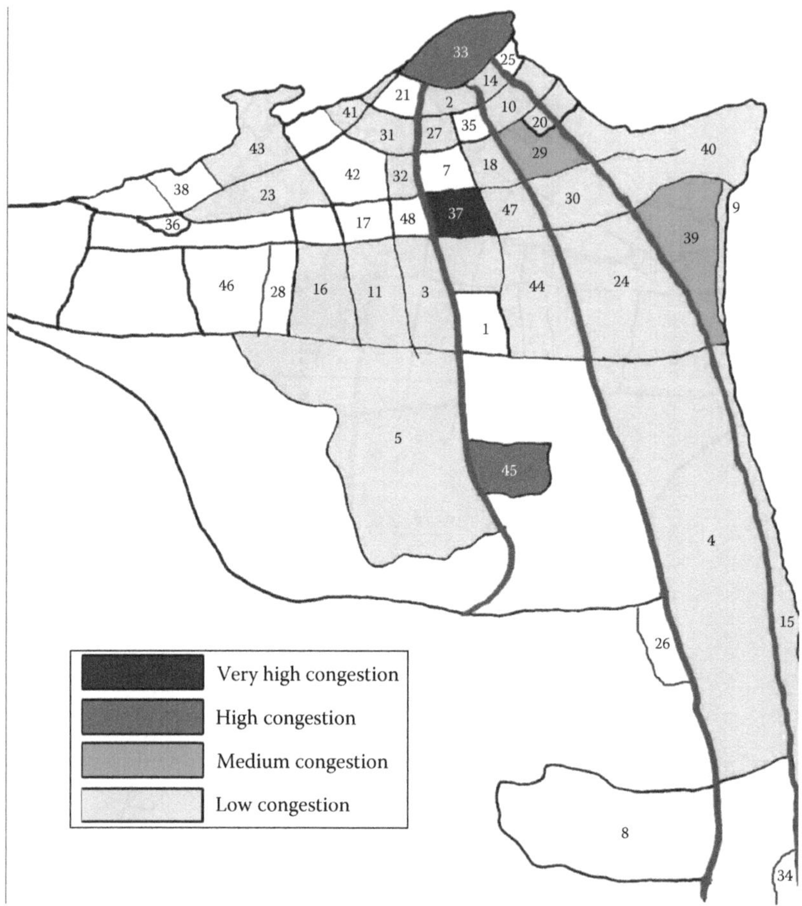

Figure 6.10 Weekdays, 12–4 p.m.

2. During this time period, Kuwait City (33) is classified as a highly congested zone. It is where employees are either leaving work or heading to work for another shift. In addition, Subhan (45), which is dedicated to employees working in the oil sector and factories, is also highly congested.
3. Hawalli (29) and Salwa and Rumaithiya (39) fall in the category of medium-congested areas mainly because students and teachers leave their schools.

Figure 6.11 shows the routes during weekdays from 4 to 6 p.m. During this time period, residents are most likely at home. This explains why residential areas in the northern part of Kuwait are congested. Road

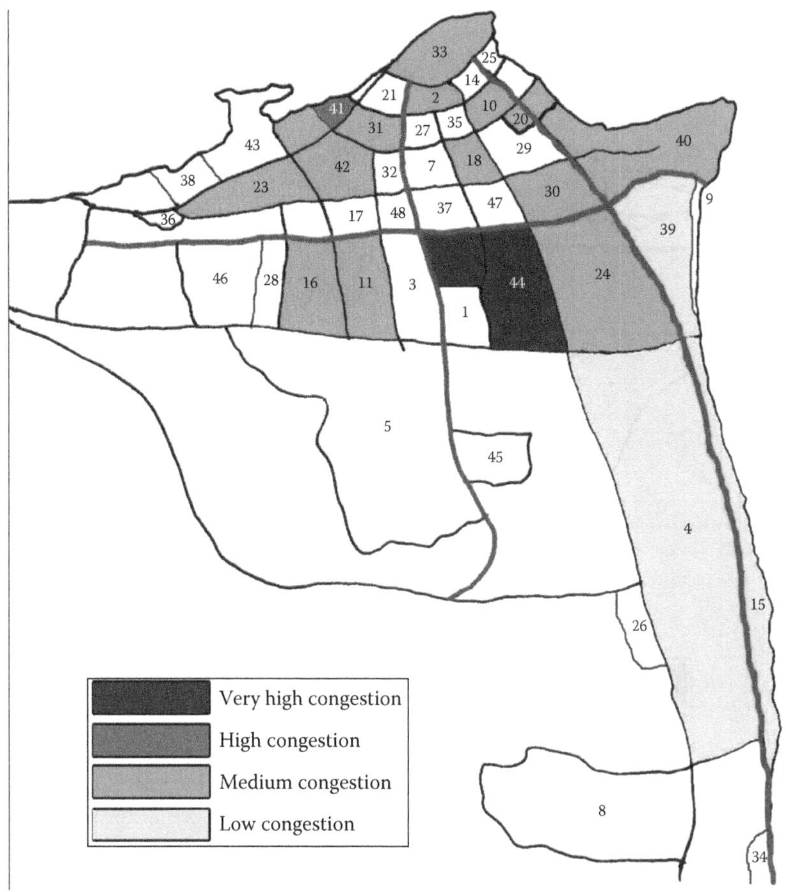

Figure 6.11 Weekdays, 4–6 p.m.

30 has a congestion level of 31.88%, followed by the 5th Ring Road with 14.49%, and finally road 50 with only 9%.

Congested routes represented in Figure 6.12 are the 4th and 5th Ring Roads both with 16.5% congestion and Fahaheel Motorway (30) with 14.56%. The reason behind heavy congestion in the 4th and 5th Ring Roads is because they both lead to Salmiya, which is a very highly congested zone (40):

1. The highest busy zone from 6 to 11 p.m. is Salmiya (40). During this time, people go to cafes, restaurants, and malls located in this area.
2. The zones with high congestion are Kuwait City (33) and Qortuba (37), which are consistent for the previous time periods.

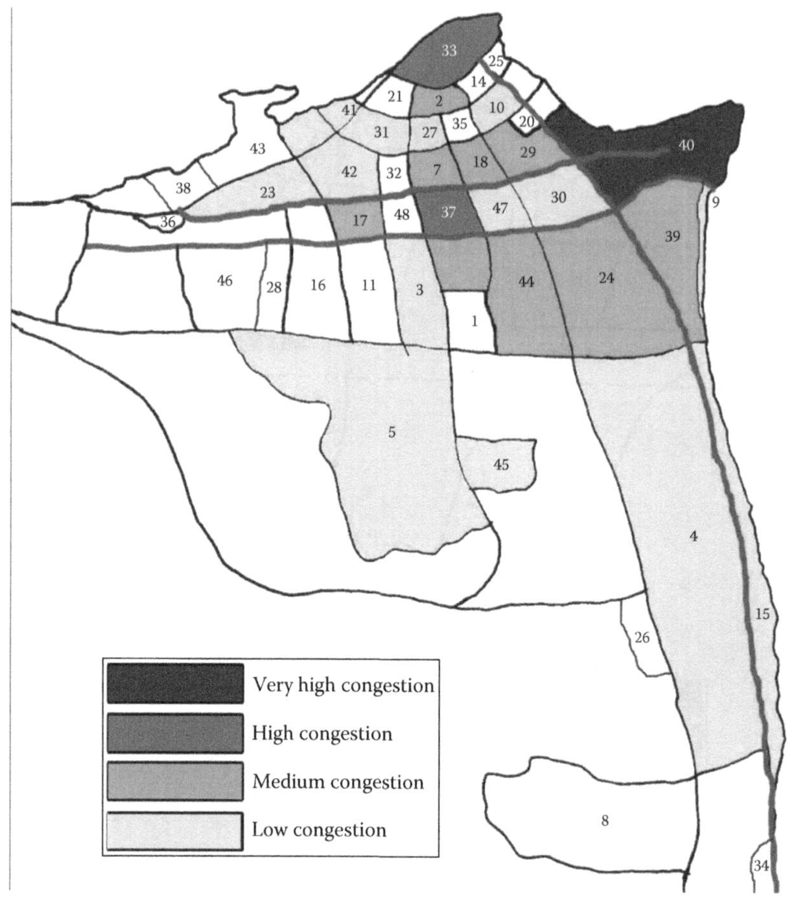

Figure 6.12 Weekdays, 6–11 p.m.

After the analysis of the weekdays, the results of the weekends are reviewed, as shown in Figure 6.13.

Congested routes represented in Figure 6.13 are the 5th and the 6th Ring Roads with 33.33% each followed by Damascus Street and Fahaheel Motorway with 16.67% each:

1. The zones classified as highly congested are Salmiya (40), containing cafes, restaurants, and shopping malls, as well as Al-Rai (17), where The Avenues mall is located.
2. The zone ranked high in congestion is a residential zone (15), consisting of Al-Messila, Abu Al-Hassani, Mahboula, Fintas, Abu Halifa, Al-Mangaf, and Fahaheel, where several restaurants are located along the coastline.

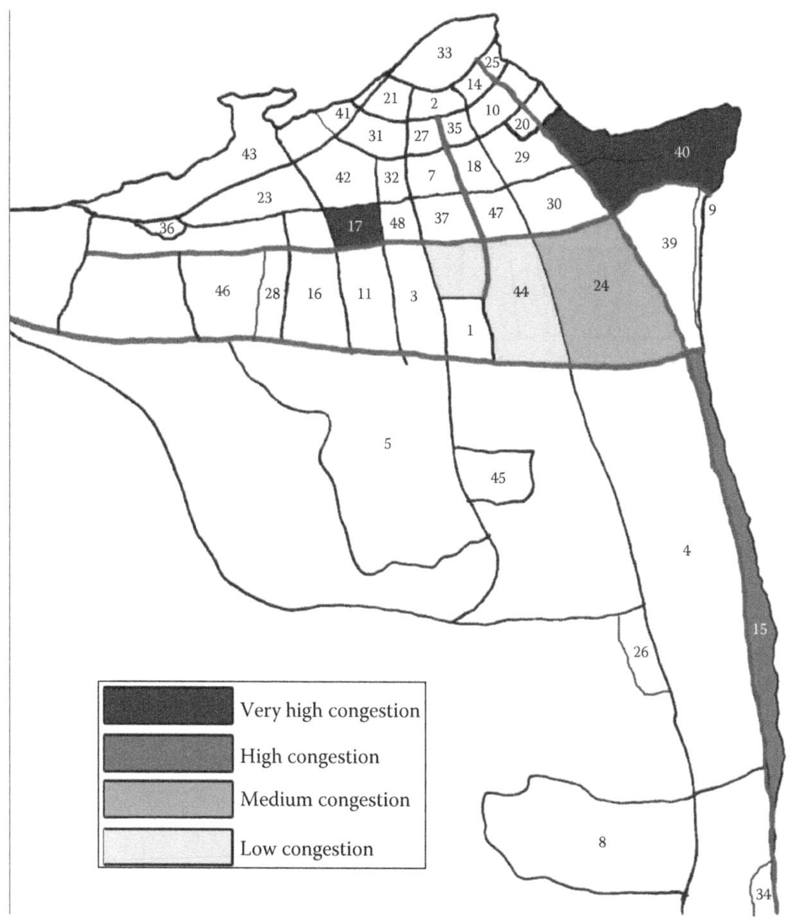

Figure 6.13 Weekend, 6–8 a.m.

Congested routes represented in Figure 6.14 are ranked starting with the 5th Ring Road with 30.30% congestion and the 4th Ring Road with 16.67%, both leading to very highly congested and highly congested zones, respectively. The third congested road is Fahaheel Motorway (30) with 15.15%:

1. The zones categorized with very high congestion are as follows:
   a. Salmiya (40), where cafes, restaurants, malls, and other activity centers are located.
   b. Qortuba (37), which is a residential area where people have family gatherings during the weekends.

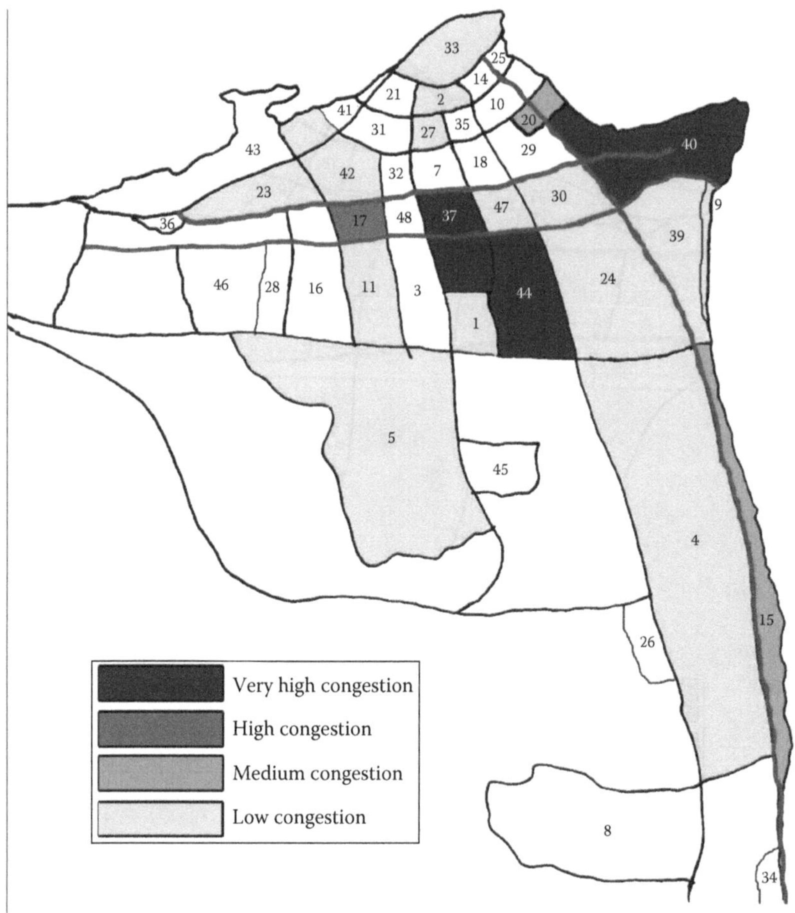

Figure 6.14 Weekend, 8 a.m.–12 p.m.

  c. South Surra (44), which is a residential area and also contains one of Kuwait's main attraction points, which is 360 Mall.

2. The zone classed as highly congested is Al-Rai (17), where one of Kuwait's main attraction points—The Avenues mall—is located.
3. Medium congested zones are (15) containing Al-Messila, Abu Al-Hassani, Mahboula, Fintas, Abu Halifa, Al-Mangaf, and Fahaheel. Also, Al-Sha'ab (20) includes restaurants, a theme park, and several complex buildings with stores, clinics, and beauty salons.

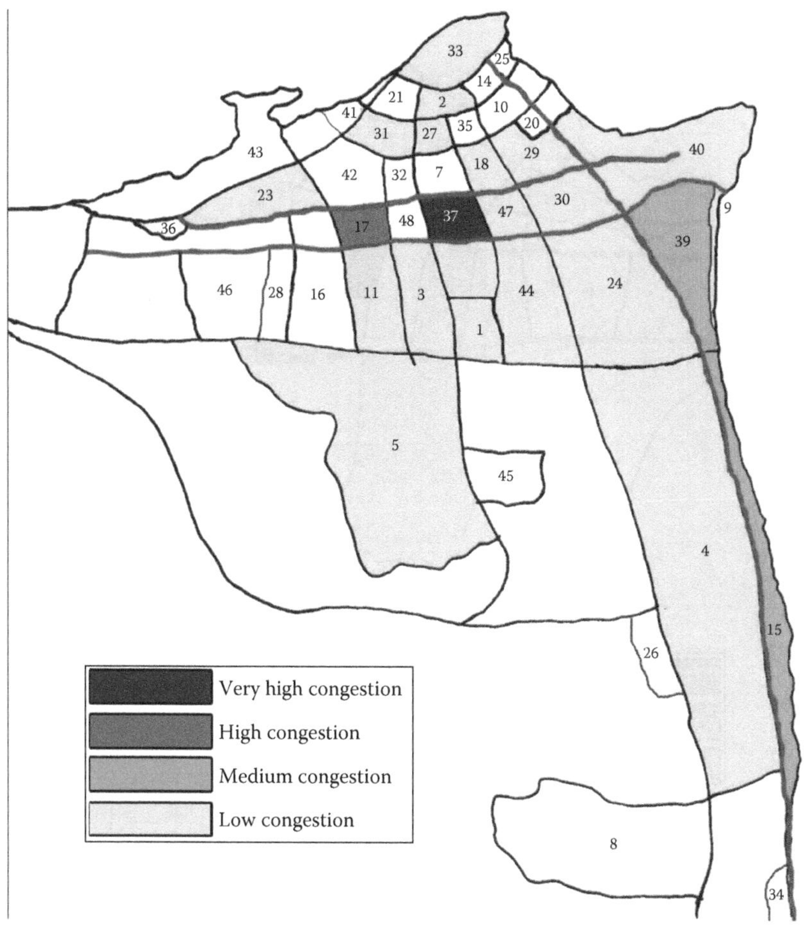

Figure 6.15 Weekend, 12–4 p.m.

Figure 6.15 shows the traffic during weekends from 12 to 4 p.m. For this time period, the congestion level for both routes and zones is still consistent and similar to the previous period of 8 a.m.–12 p.m., except for Salmiya (40) and South Surra (44), which shifted from very highly congested zones to low-congested ones.

The routes during weekends from 4 till 6 p.m. are shown in Figure 6.16. Congested routes represented in this time frame are the 5th Ring Road with 20.34% congestion compared with the total activity level of all roads, followed by the 6th Ring Road with 15.25% and Damascus Street with 13.56%.

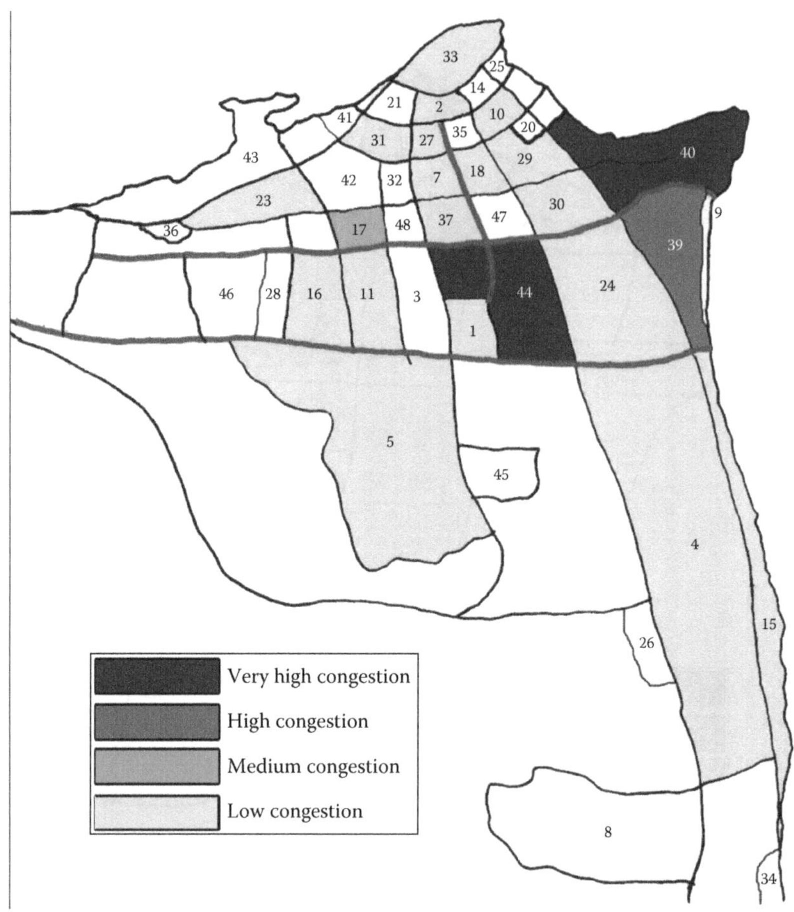

Figure 6.16 Weekend, 4–6 p.m.

1. The zones classified as highly congested are Salmiya (40), which is a main attraction point for several family activities, and South Surra (44), where 360 Mall is located.
2. The zone categorized as highly congested is (39) Rumaithiya and Salwa. These are residential areas located near the sea and are an intermediate path that people take to go to the beach.
3. The moderately busy zone is Al-Rai (17), where The Avenues mall is located.

Figure 6.17 shows the routes during weekends from 6 till 11 p.m. Congested routes represented in the figure are the same as the previous time period of 4–6 p.m. However, the congestion level increases

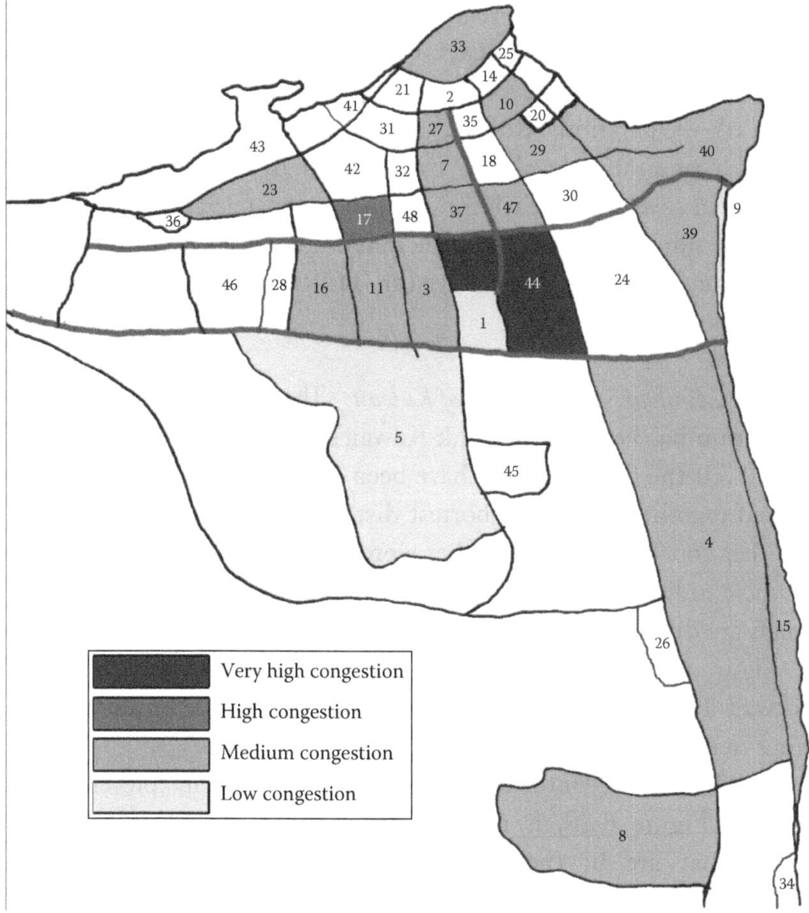

Figure 6.17 Weekend, 6–11 p.m.

to 29.20% for the 5th Ring Road, decreases to 14.16% for the 6th Ring Road, and increases to 14.16% for Damascus Street.

During this time period, the two most congested areas are Al-Rai (17) and South Surra (44), where The Avenues and 360 Mall are located, respectively.

6.4.2 Simulation and Improvement

With the complete data of Kuwait's network, such as the distance from one zone to the other and the capacity of each segment of the road, collected, a simulation must be created to arrive at a plan to handle the traffic: the shortest routes taken and the routes that can handle the most number of cars. Furthermore, improvements can be suggested. The software used to model Kuwait's network is TORA (TAHA—Operations Research: An Introduction), an operations research program, developed by Dr. Hamdy Taha. The data are entered in several files each for distance and capacity, depending on both the number of lanes and the maximum speed. Figure 6.18 is a representation of the data entered in TORA, which corresponds to Kuwait's roads capacity.

6.4.2.1 Minimal Spanning Tree of Kuwait The minimal spanning tree algorithm has been used to link Kuwait's network using the shortest roads. All the divided zones have been taken into consideration and linked together using the shortest distances. Each zone is linked to another zone separately. In other words, one highway is not taken as a whole to link all its neighboring areas; however, highways and ring roads are divided into segments.

Figure 6.19 shows the shortest routes linking all the 48 zones of Kuwait. Figures 6.19 through 6.30 show, in detail, the minimal spanning tree of Kuwait.

For Fahaheel Motorway (30), a segment of it, presented in gray in Figure 6.20, is taken as the shortest link between large zones that are in the south of Kuwait. The zones connected together with this shortest link are zone 34 (Mina Abdullah), zone 15 (Al-Messila + Abu Al-Hassani + Mahboula + Fintas + Abu Halifa + Al-Mangaf + Fahaheel), zone 4 (Adan + Qusoor + Qurain + MubarakAl-Kabeer + SabahAl-Salem + Egaila + Al-Rigga + Sabahiya), zone 39 (Rumaithiya + Salwa), and zone 24 (Bayan + Mishref).

King Fahad Motorway (40) starts at Kuwait City and ends at the Kuwait/Saudi Arabia borders. For the areas in the south of Kuwait, Fahaheel Motorway (30), as shown in Figure 6.21, is the shortest route linking them. In the case of the areas up north closer to Kuwait City, King Fahad (40) is the shortest route. The zones linked by road 40 are Abdullah Al-Salem (2), Al-Mansouriyah and Dasma (14), Nuzha (35), Al-Daiya and Qadsiya (10), Al-Rawda (18), and Hawalli (29).

The shortest road linking the residential areas Faiha (27), Nuzha (35), Al-Adailiya (7), Al-Rawda (18), Qortuba (37), Surra (47), and South Surra (44) is Damascus Street, shown in gray in Figure 6.22,

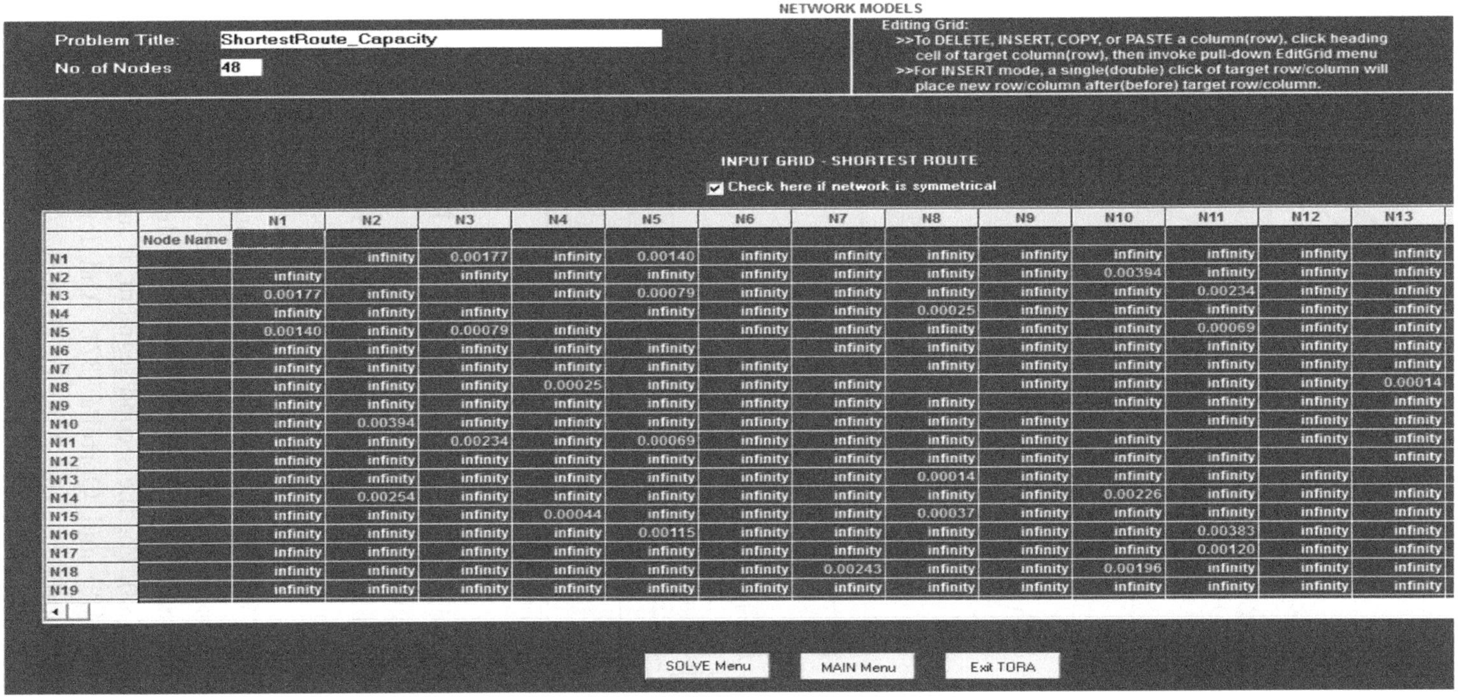

NETWORK MODELS

Problem Title: ShortestRoute_Capacity

No. of Nodes 48

Editing Grid:
>>To DELETE, INSERT, COPY, or PASTE a column(row), click heading cell of target column(row), then invoke pull-down EditGrid menu
>>For INSERT mode, a single(double) click of target row/column will place new row/column after(before) target row/column.

INPUT GRID - SHORTEST ROUTE

☑ Check here if network is symmetrical

	Node Name	N1	N2	N3	N4	N5	N6	N7	N8	N9	N10	N11	N12	N13
N1			infinity	0.00177	infinity	0.00140	infinity	infinity	infinity	infinity	infinity	infinity	infinity	infinity
N2		infinity		infinity	infinity	infinity	infinity	infinity	infinity	infinity	0.00394	infinity	infinity	infinity
N3		0.00177	infinity		infinity	0.00079	infinity	infinity	infinity	infinity	infinity	0.00234	infinity	infinity
N4		infinity	infinity	infinity		infinity	infinity	infinity	0.00025	infinity	infinity	infinity	infinity	infinity
N5		0.00140	infinity	0.00079	infinity		infinity	infinity	infinity	infinity	infinity	0.00069	infinity	infinity
N6		infinity	infinity	infinity	infinity	infinity		infinity	infinity	infinity	infinity	infinity	infinity	infinity
N7		infinity	infinity	infinity	infinity	infinity	infinity		infinity	infinity	infinity	infinity	infinity	infinity
N8		infinity	infinity	infinity	0.00025	infinity	infinity	infinity		infinity	infinity	infinity	infinity	0.00014
N9		infinity	infinity	infinity	infinity	infinity	infinity	infinity	infinity		infinity	infinity	infinity	infinity
N10		infinity	0.00394	infinity	infinity	infinity	infinity	infinity	infinity	infinity		infinity	infinity	infinity
N11		infinity	infinity	0.00234	infinity	0.00069	infinity	infinity	infinity	infinity	infinity		infinity	infinity
N12		infinity	infinity	infinity	infinity	infinity	infinity	infinity	infinity	infinity	infinity	infinity		infinity
N13		infinity	infinity	infinity	infinity	infinity	infinity	infinity	0.00014	infinity	infinity	infinity	infinity	
N14		infinity	0.00254	infinity	infinity	infinity	infinity	infinity	infinity	infinity	0.00226	infinity	infinity	infinity
N15		infinity	infinity	infinity	0.00044	infinity	infinity	infinity	0.00037	infinity	infinity	infinity	infinity	infinity
N16		infinity	infinity	infinity	infinity	0.00115	infinity	infinity	infinity	infinity	infinity	0.00383	infinity	infinity
N17		infinity	infinity	infinity	infinity	infinity	infinity	infinity	infinity	infinity	infinity	0.00120	infinity	infinity
N18		infinity	infinity	infinity	infinity	infinity	infinity	0.00243	infinity	infinity	0.00196	infinity	infinity	infinity
N19		infinity	infinity	infinity	infinity	infinity	infinity	infinity	infinity	infinity	infinity	infinity	infinity	infinity

SOLVE Menu | MAIN Menu | Exit TORA

Figure 6.18 Capacity matrix in TORA.

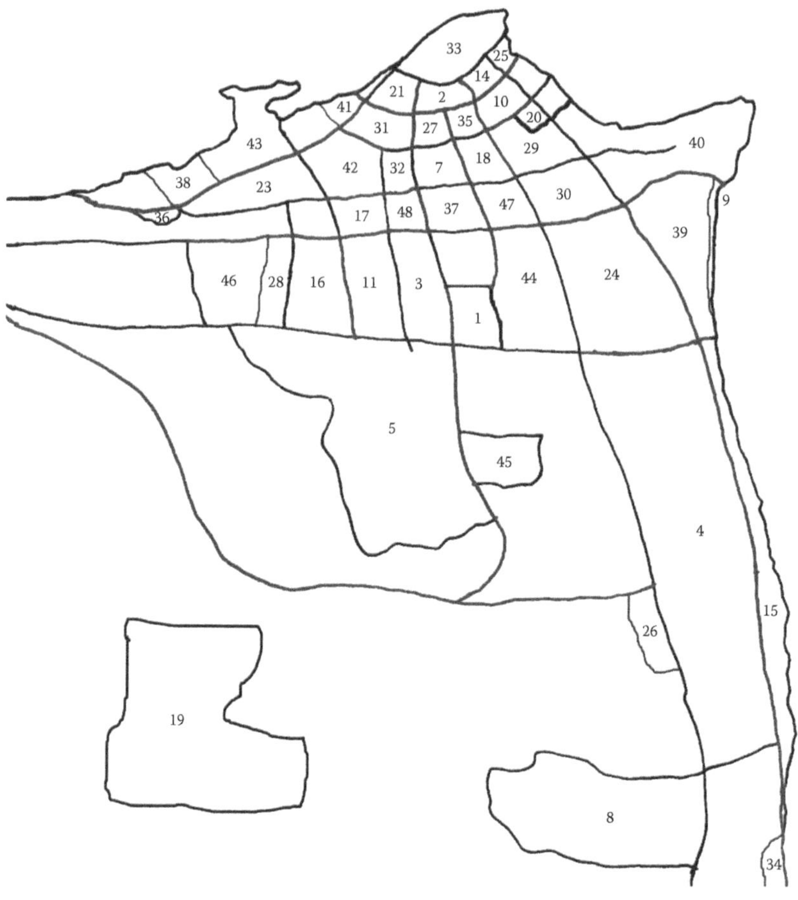

Figure 6.19 Minimal spanning tree of Kuwait.

rather than the roads King Faisal (50) and King Fahad (40) to the left and right of those areas.

King Faisal Motorway (50) starts at the 1st Ring Road and ends at the 7th Ring Road. As shown in Figure 6.23 in gray, not the entire road is considered the shortest link between the areas to the right and left of it. It is taken in segments. For the first segment in the north, the road links the residential areas Kaifan (21), Abdullah Al-Salem (2), Faiha (27), and Al-Shamiya (31) closer to Kuwait City (33). For the second segment, it links only two zones, Al-Yarmouk (48) and Qortuba (37). And for the last segment, it links Al-Zahra (1), the Airport District, Al-Dajeej, Mubarak Al-Abdullah (5), and Subhan (45).

Figure 6.20 Fahaheel Motorway in the minimal spanning tree.

As highlighted in Figure 6.24, for Al-Ghazalli road (6), only one link is available between zones 11 (Al-Farwaniya) and 16 (Al-Rabiya + Al-Rahab + Ishbilya + Ardiya).

Al-Jahra road (80) goes through all areas of Shuwaikh with the shortest distance as highlighted in gray in Figure 6.25.

In Figure 6.26, the 2nd Ring Road is taken as a whole as the shortest road linking the zones above and below it.

The 3rd Ring Road only links zone 10 (Al-Daiya + Qadsiya) to zones 20 (Al-Sha'ab) and 29 (Hawalli), and zone 31 (Kaifan) to zone 42 (Shuwaikh B [Industrial]), as shown in Figure 6.27.

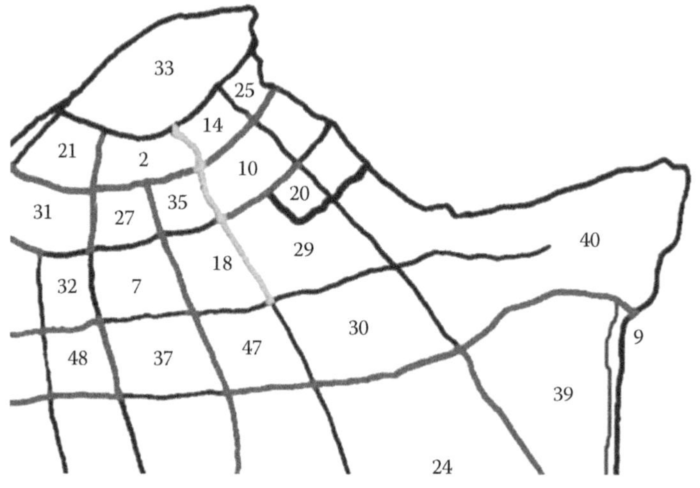

Figure 6.21 King Fahad motorway in the minimal spanning tree.

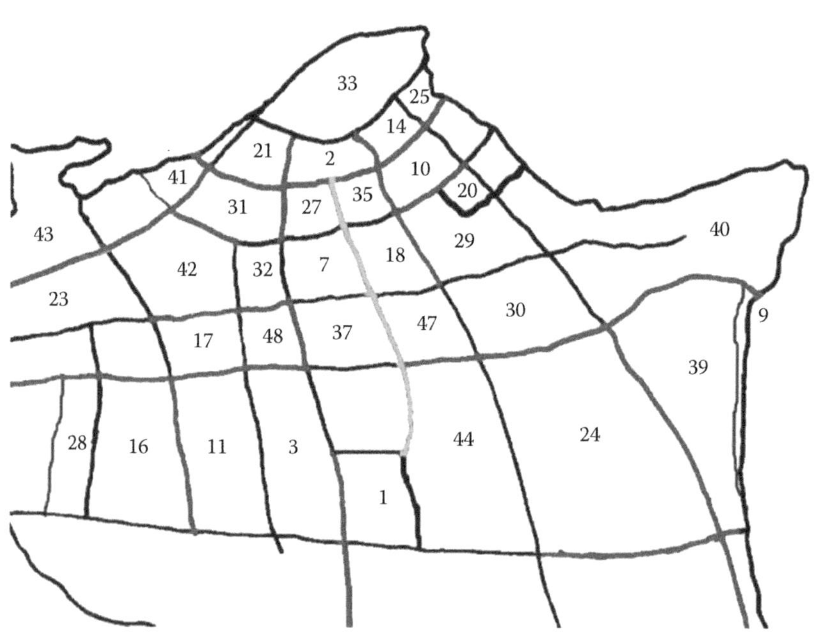

Figure 6.22 Damascus Street in the minimal spanning tree.

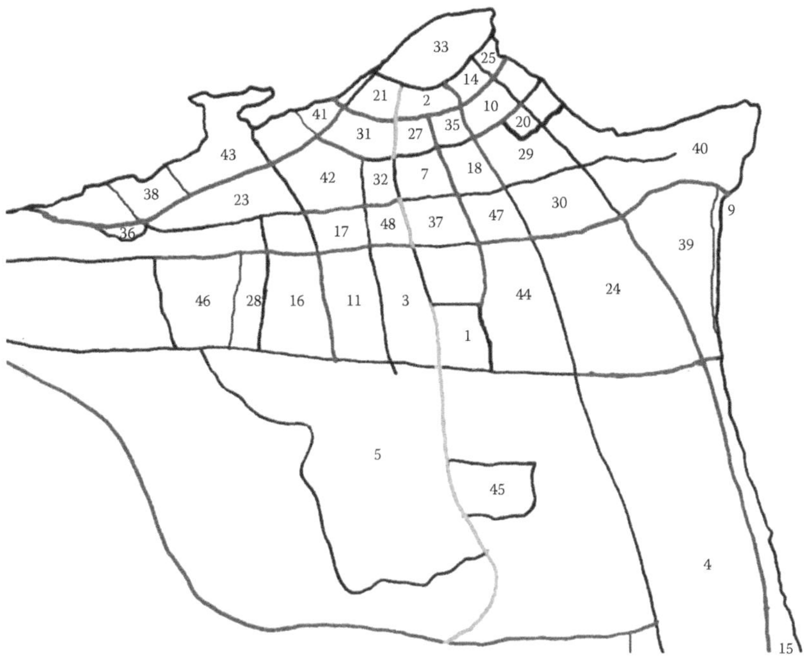

Figure 6.23 King Faisal Motorway in the minimal spanning tree.

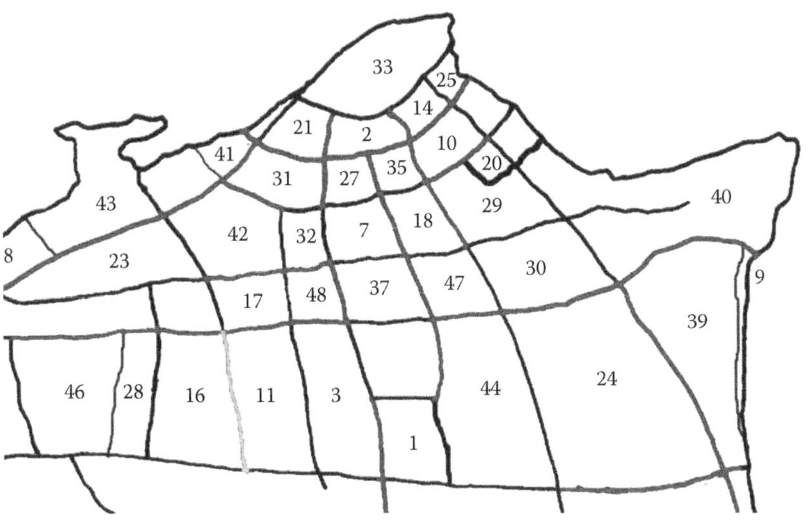

Figure 6.24 Al-Ghazalli road in the minimal spanning tree.

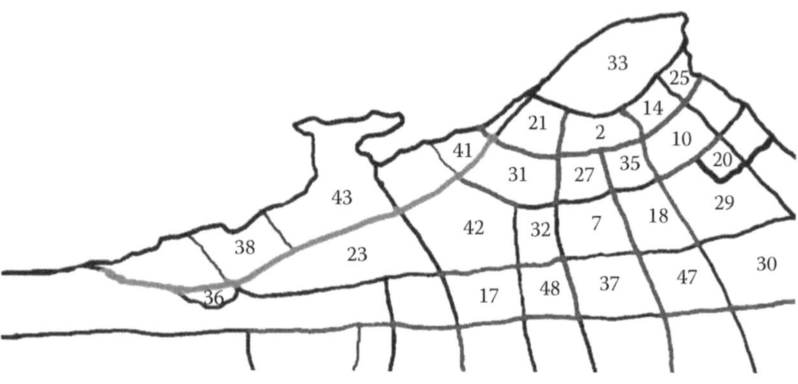

Figure 6.25 Al-Jahra road in the minimal spanning tree.

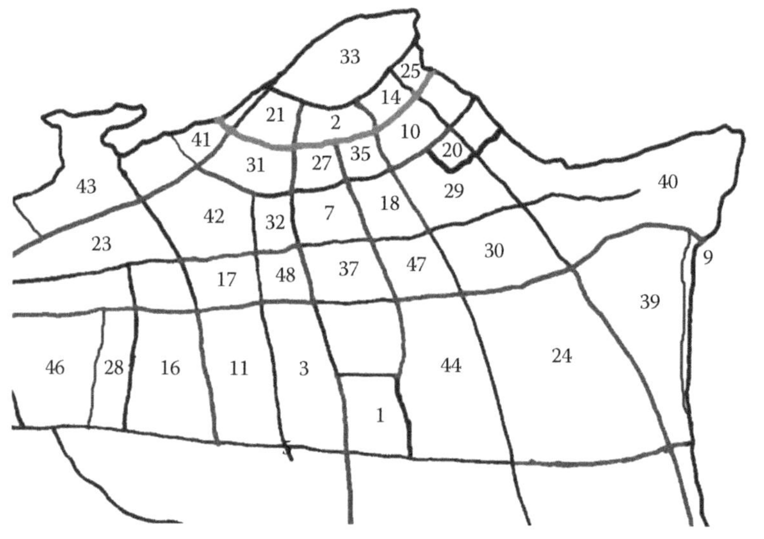

Figure 6.26 The 2nd Ring Road in the minimal spanning tree.

For the zones above the 3rd Ring Road, the 2nd Ring Road is considered the shortest road.

Damascus Street is the shortest road linking the zones below the 3rd Ring Road.

Figure 6.28 shows that one segment of the 4th Ring Road is considered the shortest link between zones 42 (Shuwaikh B [Industrial]), 32 (Khaldiya), 48 (Yarmouk), and 17 (Al-Rai).

Figure 6.29 shows the 5th Ring Road. A large segment of the 5th Ring Road is considered in Kuwait's minimal spanning tree, except

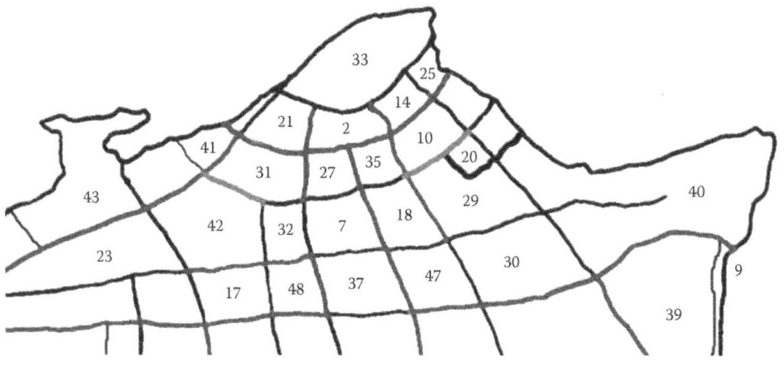

Figure 6.27 The 3rd Ring Road in the minimal spanning tree.

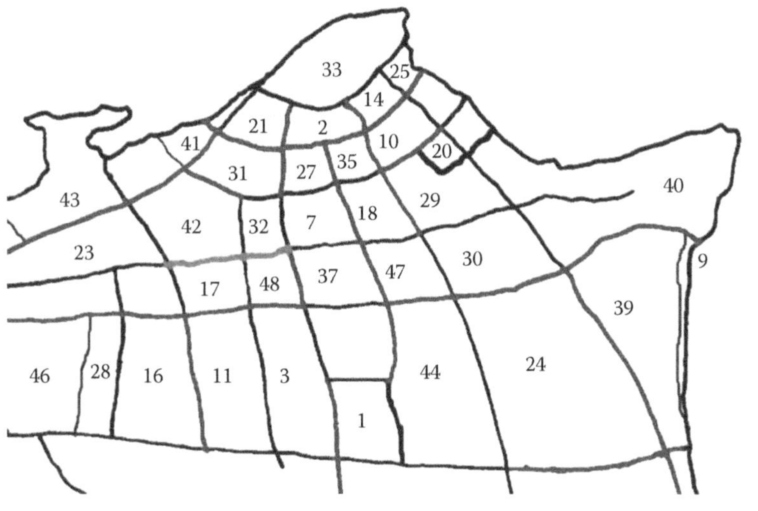

Figure 6.28 The 4th Ring Road in the minimal spanning tree.

for the link between Qortuba (37) and South Surra (44) because Damascus Street is a shorter road between these two areas.

Figure 6.30 shows that the shortest link in the 6th Ring Road is between zones 24 (Bayan + Mishref), 39 (Salwa + Rumaithiya), 15 (Al-Messila + Abu Al-Hassani + Mahboula + Fintas + Abu Halifa + Al-Mangaf + Fahaheel), and 4 (Adan + Qusoor + Qurain + Mubarak Al-Kabeer + Sabah Al-Salem + Egaila + Al-Rigga + Sabahiya).

As shown in Figure 6.31, the 7th Ring Road starts at zone 4 (Adan + Qusoor + Qurain + Mubarak Al-Kabeer + Sabah Al-Salem + Egaila + Al-Rigga + Sabahiya) and extends to the west of Kuwait, and it is the shortest route for these areas.

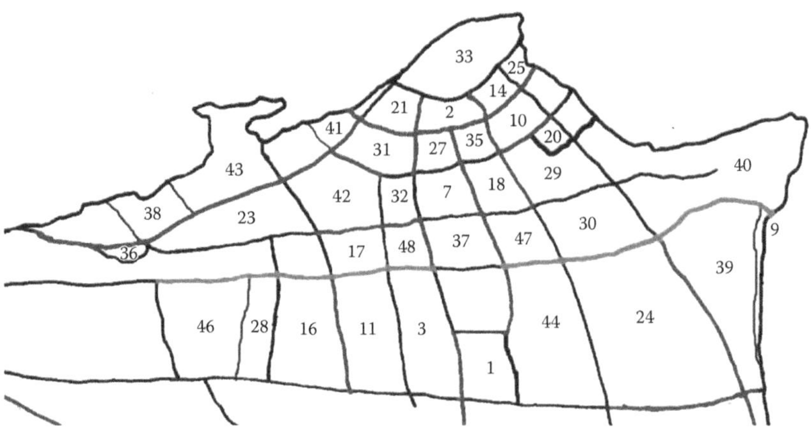

Figure 6.29 The 5th Ring Road in the minimal spanning tree.

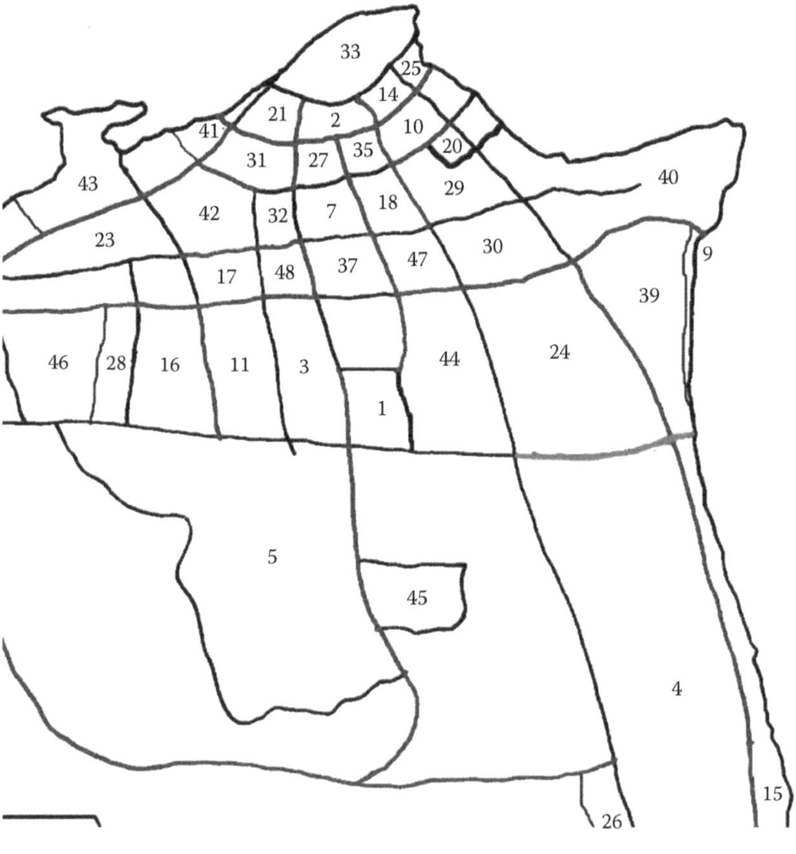

Figure 6.30 The 6th Ring Road in the minimal spanning tree.

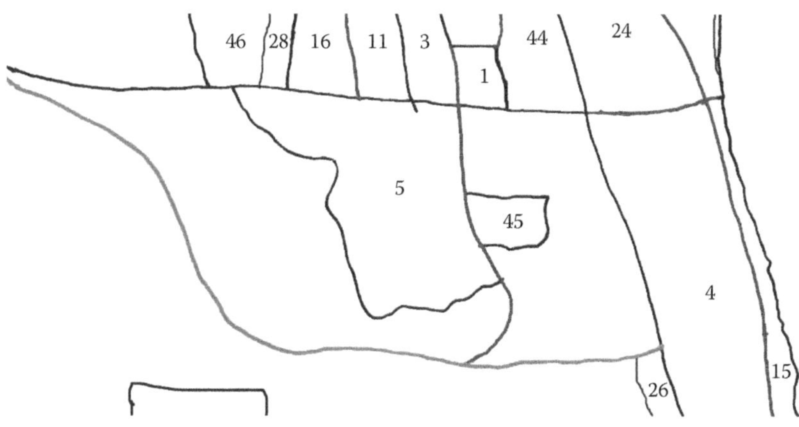

Figure 6.31 The 7th Ring Road in the minimal spanning tree.

The following description links the minimal spanning tree of Kuwait with the busy roads of the survey data.

During the two periods of 6–8 a.m. and 8 a.m.–12 p.m., the congested routes—Fahaheel Motorway (30), King Faisal Motorway (50), and the 4th Ring Road—remain the same. According to the minimal spanning tree shown in Figure 6.32, the three busy roads in those time periods that are also the shortest links between each zone in Kuwait are as follows:

1. Fahaheel Motorway (30) linking zones Rumaithiya and Salwa (39), where many private schools are located; Bayan and Mishref (24), where private universities and colleges are located; Adan, Qusoor, Qurain, Mubarak Al-Kabeer, Sabah Al-Salem, Egaila, Al-Rigga, and Sabahiya (4), which are large residential areas with private schools and universities. Also, people in those areas use the road (30) to go to their work up north. Finally, Al-Messila, Abu Al-Hassani, Mahboula, Fintas, Abu Halifa, Al-Mangaf, and Fahaheel (15), where restaurants and cafes are located by the sea.
2. King Faisal motorway—road 50—linking Subhan (45), the airport, Abdullah Al-Mubarak, and Jleeb Shuyookh (5) to Al-Zahra (1) and linking the zones close to Kuwait City (33) is congested because people tend to take the shortest route.

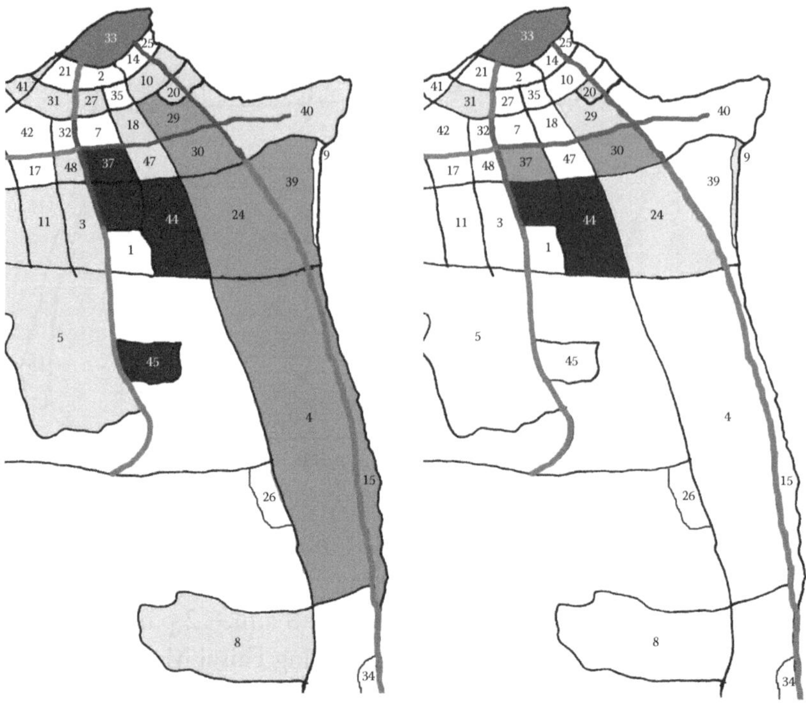

Figure 6.32 Comparison of shortest routes to busy routes from 6 a.m. to 12 p.m.

6.4.2.2 Scenarios and Improvements The algorithm applied in this section for the network model of Kuwait is the shortest route method to

1. Find the shortest route in terms of distance (km) from and to every zone
2. Find the routes with the highest capacities that link one zone to the other

Doing so will aid in further analysis of all the areas and roads of Kuwait and their optimal linkages. The simulation results were compared with the cluster analysis results taken from the survey data.

For each time period, the busiest zones and roads shown in the cluster analysis maps given in Section 6.4.1.2 have been taken in order to compare the optimal results with the real-life situation.

From 6 to 8 a.m., the zone ranked very high in congestion is South Surra (44). It is a residential area and a destination point for

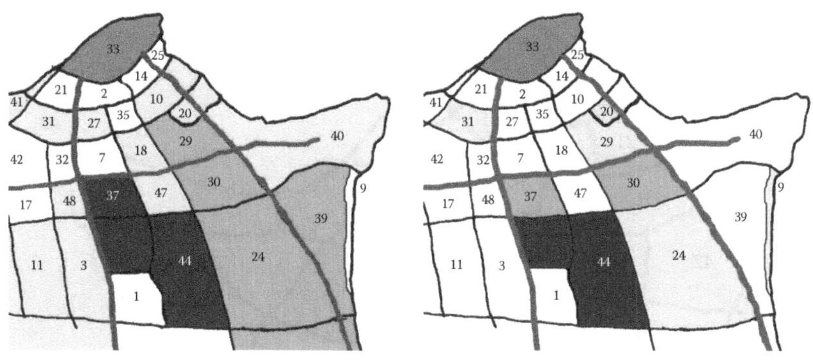

Figure 6.33 Congestion on weekdays—6–8 a.m. and 8 a.m.–12 p.m.

the employees and clients of the Ministry of Electricity and Water, the Public Authority for Civil Information, and other organizations such as the International Islamic Charitable Organization. During this time, Kuwait City (33) is considered to be a highly congested area. It is the target of many ministries such as the Ministry of Higher Education, banks such as the Industrial Bank of Kuwait, and many private companies and organizations. These two zones remain congested for the time period of 8 a.m.–12 p.m.

As for the routes, the busiest highways leading to the most congested zones (44 and 33) during the two mentioned time periods are Fahaheel Motorway (30) and King Faisal Motorway (50).

When the shortest route algorithm to the network is applied, the shortest route from zone 44 to zone 33 is King Faisal Motorway (50), with 6.83 km. The route with the highest capacity linking these two zones is King Fahad Motorway (40), as shown in Figure 6.33.

6.4.2.2.1 Scenario 1 As shown in Figure 6.33, King Fahad Motorway (40) is not congested; therefore, people tend to take the shortest route from South Surra (44) to Kuwait City (33) via King Faisal Motorway (50). This has led to the case of increasing the capacity of King Faisal Motorway (50) since it is the shortest link and the frequently used road.

To increase the capacity of road 50, a lane was added. The road originally consisted of three lanes and was increased to four. When the new capacity of the road was simulated, the road turned out to be

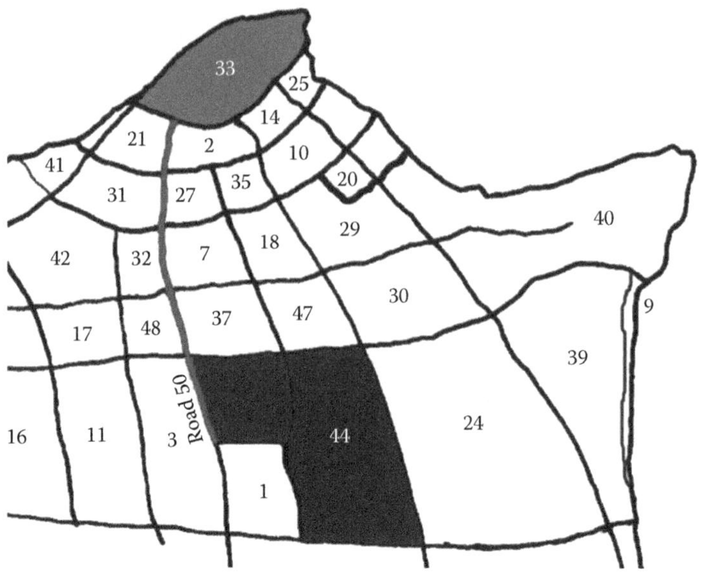

Figure 6.34 Optimal route.

Table 6.6 Capacity Increase of Road 50

THREE-LANE CAPACITY	FOUR-LANE CAPACITY	PERCENT INCREASE
3328	4156	19.92%

the optimal route, linking zone 33–44 in terms of distance and capacity, as shown in Figure 6.34. This improvement will also reduce the congestion level of residential areas, such as Qortuba (37).

Table 6.6 compares the capacity increase of the segment between zones 44 and 33 before and after the improvement.

6.4.2.2.2 Scenario 2 In case of a blockage, such as that caused by an accident, where road 50 cannot handle any more cars, route 40 would be the best alternative in terms of capacity. This case is most likely to happen since road 50 is busy during 6 a.m.–12 p.m., as shown in Figure 6.35.

6.4.2.2.3 Scenario 3 Another scenario taken is Salmiya being a destination point. In weekdays, from 6 to 11 p.m., Salmiya is highly congested; therefore, the routes leading to it are also busy. According

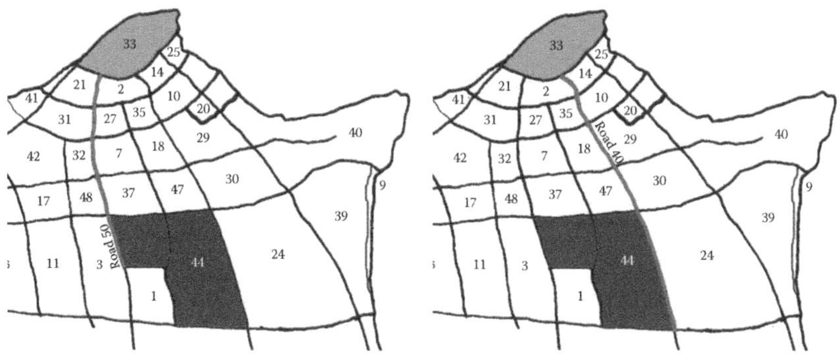

Figure 6.35 Congested route and alternative route.

to the results of the surveys, the two most congested routes leading to Salmiya are the 4th and 5th Ring Roads.

Taking a highly congested residential area in between the 4th Ring Road and the 5th Ring Road—Qortuba (37)—and applying the shortest route algorithm for determining the road with the shortest distance and highest capacity to reach this destination.

As represented in Figure 6.36, the optimal route to take from zone (37) to (40) is the 5th Ring Road in terms of the highest capacity and both the 4th and 5th Ring Roads in terms of the shortest distance of 4.65 km.

To reduce the congestion on the 5th Ring Road, a new lane was added to the 4th Ring Road, increasing its capacity. In this way, the congestion would be distributed among the 4th and the 5th Ring Roads.

In simulation, it is found that the optimal route to take from Qortuba (37) to Salmiya (40) is the 4th Ring Road, as shown in Figure 6.37.

Table 6.7 shows the increase in capacity after the improvement.

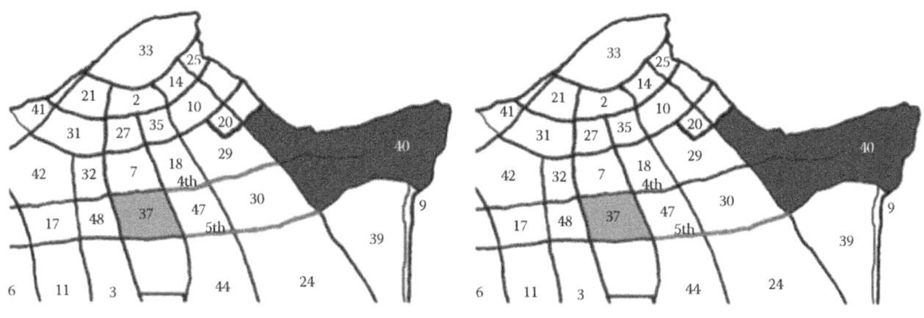

Figure 6.36 Roads with the shortest distance and the highest capacity.

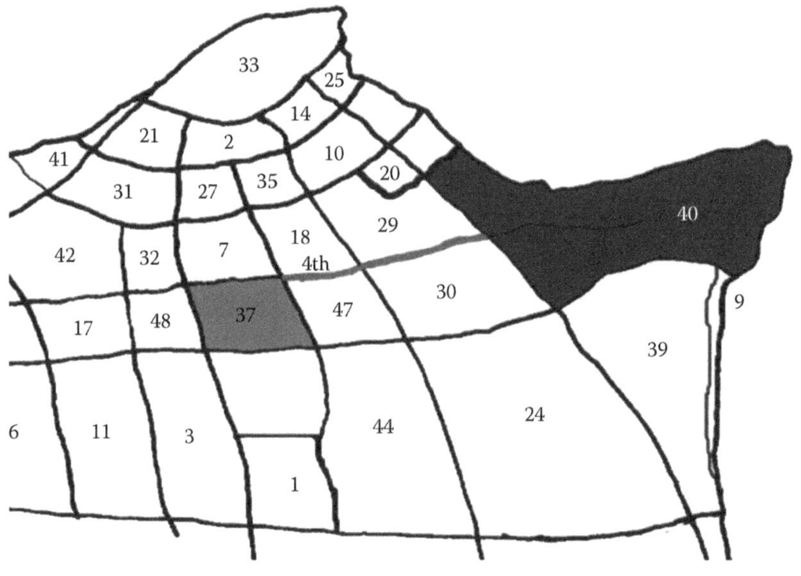

Figure 6.37 Optimal route from (37) to (40).

Table 6.7 Capacity Increase of 4th Ring Road

THREE-LANE CAPACITY	FOUR-LANE CAPACITY	PERCENT INCREASE
1295	1517	14.63%

The previous scenarios represented weekday behaviors. As for the weekend, different behaviors were noticed.

As shown in Figure 6.38, Salmiya (40), South Surra (44), and Qortuba (37) were considered very highly congested areas during the 8 a.m.–12 p.m. time period. Salmiya is a popular destination point, which includes shopping malls, restaurants, cafes, and other entertainment destinations. Qortuba and South Surra are residential areas where family gatherings occur, especially during the weekend. Al-Rai (17) is another destination for those going for shopping, automobile companies, and car garages. One of Kuwait's most famous attraction points, The Avenues mall, is located in this zone. According to the map in Figure 6.38, The Avenues is considered a highly congested area.

The two roads leading to Salmiya (40) and Al-Rai (17) are the 4th and 5th Ring Roads. As represented in Figure 6.38, these two roads are highlighted, meaning that they are busy during this time period.

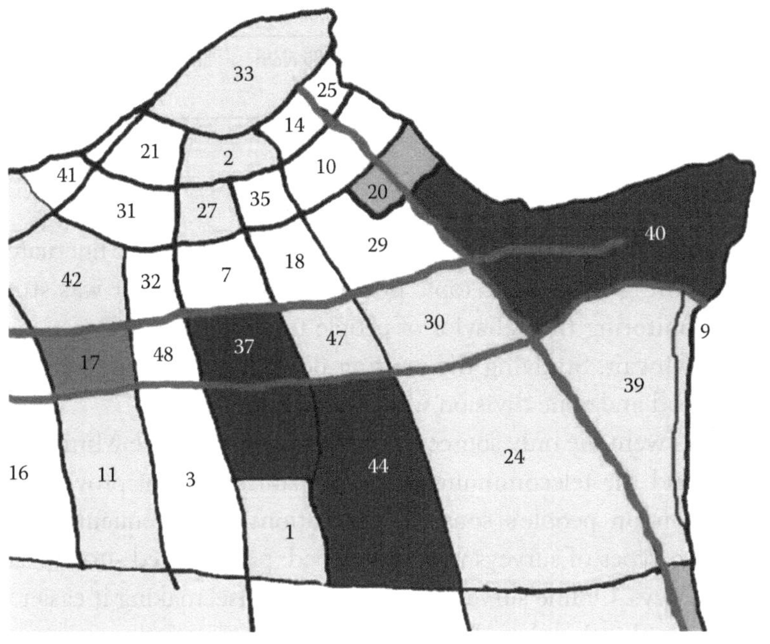

Figure 6.38 Destination points Al-Rai (17) and Salmiya (40).

Simulation was done to find the best route leading to Al-Rai (17). According to capacity, the optimal route leading to zone 17 was the 5th Ring Road, and according to distance, the 4th Ring Road is a shorter route, 3.7 km, to go to zone 17. In order to improve this situation and reduce the overcrowding of the 5th Ring Road, an increase in the 4th Ring Road capacity is needed. The maximum speed was increased, making it 100 km/h instead of 80 km/h. After simulating the new scenario, the optimal road leading to zone 17 is still the 5th Ring Road in terms of capacity. However, as shown previously in scenario 3, the best route leading to Salmiya (40) is the 4th Ring Road after the improvement in terms of distance and capacity.

Table 6.8 shows the capacity changes between Qortuba (37) and Salmiya (40), whereas Table 6.9 shows the capacity changes between Qortuba (37) and Al-Rai (17).

Table 6.8 Capacity Increase of the 4th Ring Road Leading to Salmiya (40)

MAXIMUM SPEED 80 KM/H	MAXIMUM SPEED 100 KM/H	PERCENT INCREASE
1262	1323	4.61%

Table 6.9 Capacity Increase of the 4th Ring Road Leading to Al-Rai (17)

MAXIMUM SPEED 80 KM/H	MAXIMUM SPEED 100 KM/H	PERCENT INCREASE
849	883	3.85%

6.5 Conclusion

With the increase in population and emigration, the traffic fluctuates, only becoming worse. The topic presented in this chapter was studied by monitoring the behavior of people to understand the trends of roads in Kuwait. Studying the existing flow of traffic along with the current road and zone division was fundamental.

Surveys were the only source for gathering data, as the Ministry of Interior and the telecommunication companies did not provide any information on people's sources, destinations, and frequently used roads. Two types of surveys were developed: paper-based surveys and online surveys. Online surveys had faster response, making it easier to distribute and gather data. From the surveys, time periods were developed. Specifically, five time periods divided the day; time periods were directed for both weekdays and weekends. Constructing matrices, including the zones along with the routes, was next the step, after the time period division. The matrices were different for every time period of the day and for both weekends and weekdays. They were the tools used to develop the cluster analysis maps. Those maps were developed using a statistical technique called hierarchical cluster analysis, which grouped the level of congestion in a hierarchical form, ranking from the highest to the lowest. Finally, a representation of the road congestions was viewed through the weather map evolution.

Simulation helped in visualizing the existing road conditions and therefore contributed to the development of new scenarios. At first, the simulation of the existing flow of traffic was done, and then the simulation of the capacity of the roads and the distance of each road was done to detect the optimal route from one place to another. Moreover, speed limits were altered and shifted to reach the ideal road capacity to ultimately reduce traffic. Throughout the procedure, peak times were always considered regarding the congestion of roads. Thus, from the simulation of different situations, several improvements and alternatives were suggested.

The future studies on this topic could be dealt with in more detail. Since this chapter covers Kuwait's main roads and highways, inner roads of the zones were not taken into consideration. However, the addition of these inner roads, roundabouts, and intersections can be an interesting addition to the study. Also, attention will be paid to traffic lights for more precise simulation of Kuwait's existing roads as they also cause delays. This will result in a microscopic view of the traffic in Kuwait.

Appendix 6.A: Surveys

6.A.1 Paper-Based Survey

6.A.1.1 Analyzing Kuwait's Road Traffic

This survey is regarding the graduation project for industrial engineering students. We are aiming to collect data concerning the traffic load on the Kuwait roads and highways.

6.A.1.1.1 Weekdays

Gender

- Male
- Female

Age

- Less than 18
- 18–25
- 26–45
- 46 and above

Occupation

- Student
- Employee
- Retired
- Other (please specify) ____________________

Work/Class Destination (If you drop your kids to schools, please mention the areas in order)

From: ____________________ To: ____________________
From: ____________________ To: ____________________
From: ____________________ To: ____________________

Going to work or class, what route(s) do you take? (If you take more than one road, please number them by their order.)

- 1st Ring Road
- 2nd Ring Road
- 3rd Ring Road
- 4th Ring Road
- 5th Ring Road
- 6th Ring Road
- 7th Ring Road
- Damascus street
- Fahaheel Motorway
- Al-Ghazally Street (60)
- King Fahad Bin Abdulaziz (40)
- Maghreb Street
- King Faisal Motorway—Airport Road (50)
- Other (please specify) ________________

Leaving work or class, what route(s) do you take? (If you take more than one road, please number them by their order.)

- 1st Ring Road
- 2nd Ring Road
- 3rd Ring Road
- 4th Ring Road
- 5th Ring Road
- 6th Ring Road
- 7th Ring Road
- Damascus street
- Fahaheel Motorway
- Al-Ghazally Street (60)
- King Fahad Bin Abdulaziz (40)
- Maghreb Street
- King Faisal Motorway—Airport Road (50)
- Other (please specify) ________________

Time leaving the house ______ *Time of arrival to your work* ______
Time leaving work ______ *Time of arriving home* ______

6.A.1.1.2 Weekends

What is your preferred time for going out during the weekend?

- Morning
- Afternoon
- Evening

What is your preferred destination during the weekend?

From: ________________
To: ________________

From home to the desired destination, what route(s) do you take? (If you take more than one road, please number them in their order.)

- 1st Ring Road
- 2nd Ring Road
- 3rd Ring Road
- 4th Ring Road
- 5th Ring Road
- 6th Ring Road
- 7th Ring Road
- Damascus street
- Fahaheel Motorway
- Al-Ghazally Street (60)
- King Fahad Bin Abdulaziz (40)
- Maghreb Street
- King Faisal Motorway—Airport Road (50)
- Other (please specify) ____________________

What route(s) do you take going back home? (If you take more than one road, please number them in their order.)

- 1st Ring Road
- 2nd Ring Road
- 3rd Ring Road
- 4th Ring Road
- 5th Ring Road
- 6th Ring Road
- 7th Ring Road
- Damascus street
- Fahaheel Motorway
- Al-Ghazally Street (60)
- King Fahad Bin Abdulaziz (40)
- Maghreb Street
- King Faisal Motorway—Airport Road (50)
- Other (please specify) ____________________

Time leaving the house __________ *Time of arrival to destination* _____
Time leaving the destination _____ *Time of arriving home* __________

6.A.1.1.3 After Work

During weekdays, do you go to specific places regularly? If yes, please fill this page.

Kindly write down your destination below:

From: ___________________
To: ___________________

What road(s) do you take going to your desired destination? (If you take more than one road, please number them by their order.)

- 1st Ring Road
- 2nd Ring Road
- 3rd Ring Road
- 4th Ring Road
- 5th Ring Road
- 6th Ring Road
- 7th Ring Road
- Damascus street
- Fahaheel Motorway
- Al-Ghazally Street (60)
- King Fahad Bin Abdulaziz (40)
- Maghreb Street
- King Faisal Motorway—Airport Road (50)
- Other (please specify) ____________________

What road(s) do you take going back home? (If you take more than one road, please number them by their order.)

- 1st Ring Road
- 2nd Ring Road
- 3rd Ring Road
- 4th Ring Road
- 5th Ring Road
- 6th Ring Road
- 7th Ring Road
- Damascus street
- Fahaheel Motorway
- Al-Ghazally Street (60)
- King Fahad Bin Abdulaziz (40)
- Maghreb Street
- King Faisal Motorway—Airport Road (50)
- Other (please specify) ____________________

Time leaving the house __________ *Time of arrival to destination* _____
Time leaving the destination _____ *Time of arriving home* __________

Thank you for your time. ☺

6.A.2 Online-Based Survey

1. Gender

- Male
- Female

2. Age

- Less than 18
- 18 - 24
- 25 - 45
- 46 - Above

3. Occupation

- Student
- Employee
- Retired
- Other

4. Kindly state your location (area) during working days according to the time periods , eg; Bayan, Shuwaikh

6:00 - 8:00 am	
8:00 - 12:00 am	
12:00 - 4:00 pm	
4:00 - 6:00 pm	
6:00 - 11:00 pm	

5. Kindly state your location (area) during the weekend according to the time periods

6:00 - 8:00 am	
8:00 - 12:00 am	
12:00 - 4:00 pm	
4:00 - 6:00 pm	
6:00 - 11:00 pm	

Done

References

1. Kuwait Government Online. (2011). Population of Kuwait. Retrieved November 25, 2012, from http://www.e.gov.kw.
2. Ministry of Interior. (2007). Al-Mururiya magazine, issue No. 9. Retrieved November 30, 2012, from http://www.moi.gov.kw/portal/vArabic/storage/other/mjm9.pdf.
3. N.A. (2008). Sample survey questions, answers and tips. Retrieved February 28, 2013, from http://www.constantcontact.com/aka/docs/pdf/survey_sample_qa_tips.pdf.
4. Taha, H. (2010). *Operations Research: An Introduction*. 9th edn. Prentice Hall, NJ: Pearson Education.
5. Bosch, R. and Trick, M. (2005). *Search Methodologies: Introductory Tutorials in Optimization and Decision Support Technique*. Oberlin, OH: Springer.
6. Moradkhan, M.D. (2010). Multi-criterion optimization in minimum spanning trees. *Studia Informatica Universalis*, 8(2): 185–208.

7. Li, D.-L., Li, R.-W., Li, Y.-H., and Zhang, P.-J. (2009). Improved spanning tree-based genetic algorithm and its application in cost optimization of logistics dispatching system. *Mathematics in Practice and Theory* (21): 38–44.
8. Dijkstra, E.W. (1959). A note on two problems in connexion with graphs. *Numerische Mathematik*, 1(1): 269–271.
9. Teodorovic, D. (1986). *Transportation Networks*. New York: Gordon & Breach Science Publishers.
10. Chvatal, V. (1983). *Linear Programming*. New York: W.H. Freeman.
11. Godden, B. (2004). *Sample Size Formula*. Chicago, IL: Marketing Research Association.
12. Nicholas, J.G. and Lester, A.H. (2002). *Traffic and Highway Engineering*. Bill Stenquist, Books/Cole Thomson Learning.
13. Google. (2012). Google Maps. Retrieved December 9, 2012, from https://www.google.com.kw/maps/@29.1924637,47.7780801,10z?hl=en
14. Mardia, K., Kent, J., and Bibby J. (1980). *Multivariate Analysis* (*Probability and Mathematical Statistics*). London, U.K.: Academic Press Inc.

7

Modeling, Simulation, and Analysis of Production Lines in Kuwait's Petroleum Sector

SEIFEDINE KADRY, RAWAN JARAGH, REEM AL-MADY, SHAHAD SHEER, AND SHAIKHA AL-DABBOUS

Contents

7.1 Introduction

A production line is a repetitive manufacturing process in which the product passes through the same sequence of operations. Gaining a thorough understanding of a production line grants you the ease of tackling rising issues. Hence, industrial engineers seek to improve operation systems wherever they are placed. They target clear objectives throughout the stages of production.

This chapter would convey the analysis and examination of production lines to find surfacing issues in order to find ways to improve them.

7.2 Methodologies and Approaches

Throughout this research, data were collected from the oil sector in Kuwait. The data were used to discover if any enhancements can be made to improve the production rate of the company. Different methods were found to approach the project.

7.2.1 Heuristic Method

According to Groner et al. (1983), a heuristic method is a classical problem-solving method that has been used by many to speed up the process by retracing steps to prove the outcome. It starts off with an analysis and a concrete assumption as a mean for proving if the expected consequences are in fact true or not. This method relies strongly on trial-and-error and up-to-date mathematical results for gaining a step-by-step path toward anticipated results.

This method (Cortes et al., 2009) was used in a case study aimed to solve the assembly line-balancing problem of a motorcycle manufacturing company. The method worked well; however, it did not provide accurate results. Therefore, this method will be eliminated from the options at hand.

7.2.2 Mathematical Programming

Mathematical (linear) programming is one of the most popular methods for modeling with the purpose of increasing one's profitability. It easily provides accurate results when used with simple production lines containing a single queue and a few stations.

Saad et al. (2009) used such method as a tool for crude oil scheduling. Although this method may provide concrete facts, it requires a higher-level degree.

The reason for the dismissal of this method is the current lack of availability of mathematical software and tools. Moreover, although this method may provide accurate results, it would turn complicated with complex systems.

7.2.3 Computer Simulation

Kleijenen (2008) refers to computer simulation as a trial-and-error approach where systems can be modeled with regard to statistical methods. When a set of input data is ready for evaluation, placing it in a simulation software provides a simulated model where changes can be applied without the need of changing the actual environment.

This methodology was found most suitable for this project since it does not require a large financial investment and does not consume extreme periods of time. According to Carson and Maria (1997), using a simulation approach, one can manipulate the different parameters to compare between scenarios in order to seek the most suitable scenario. If optimized results were found, the improved model may be suggested to the company where they can safely implement the changes, considering that such suggestions were backed up with evidence.

This research will use the computer simulation software, Arena®. The software is user friendly and provides thorough statistical results, and the group members are familiar with the software.

7.2.4 Application of Arena

Prior to designing the project's simulation model, one needs to research the topic to gain insight on how simulation using Arena was implemented previously. Cortes et al. (2009) simulated the assembly line of a motorcycle manufacturing company in order to solve the line-balancing problem. Several scenarios were developed using Arena as a solution to the issue. They modeled the company's current assembly line and improved it using a couple of different approaches. Both approaches chosen proved to resolve the problem; however, they chose the scenario that resulted in accurate balancing of the production line with an increase in productivity.

According to Hecker et al. (2010), when planning to simulate a real-life production line, it is preferable to follow the 40-20-40 rule to ensure optimization. This rule states that 40% of the time should be dedicated to the gathering of data. The 20% that follows is for the designing and simulation of the model. The remaining 40% is assigned for the verification of the model as well as the validation. During that 40% of the time, one would start enhancing the process in order to modify and implement the changes. The aim of the journal was to present existing bakeries with an opportunity to improve their production plan in accordance to machine utilization and energy consumption. Arena was used to simulate an existing scenario; such scenario was further altered to reach beneficial results. They were able to reduce the energy consumed by three machines and found a way to save 32% of the salaries given out by decreasing the total shift time. This study is the first in the domain of petroleum sector.

Figure 7.1, in the following page, depicts a flowchart that examines the framework in the simulation procedure. The process begins with formulating a problem and ends with documenting the findings as well as the implementation.

7.3 Process Description

Ideally, one would seek to influence the production rate, a key project parameter. The following parameters have been set as a guideline when visiting a company:

- Employee utilization
- Input and output
- Waiting time
- Optimizing process flow
- Cost

One parameter will be changed, while the rest remain controlled. The influence of the chosen parameter on the production rate would be noted.

When a customer request arrives at the refinery, the company receives a customized blend request form matching the customer's specification. The process is performed by the controller unit, with the capacity of 4 employees in each 1 of the 10 units. Once the blending process has been completed, a sample is to be taken to the laboratories for testing to see whether the blend complies with the customer's request. If it does not, then the mixture goes through a process known as *intermediate blending*. This cycle continues until the lab test proves

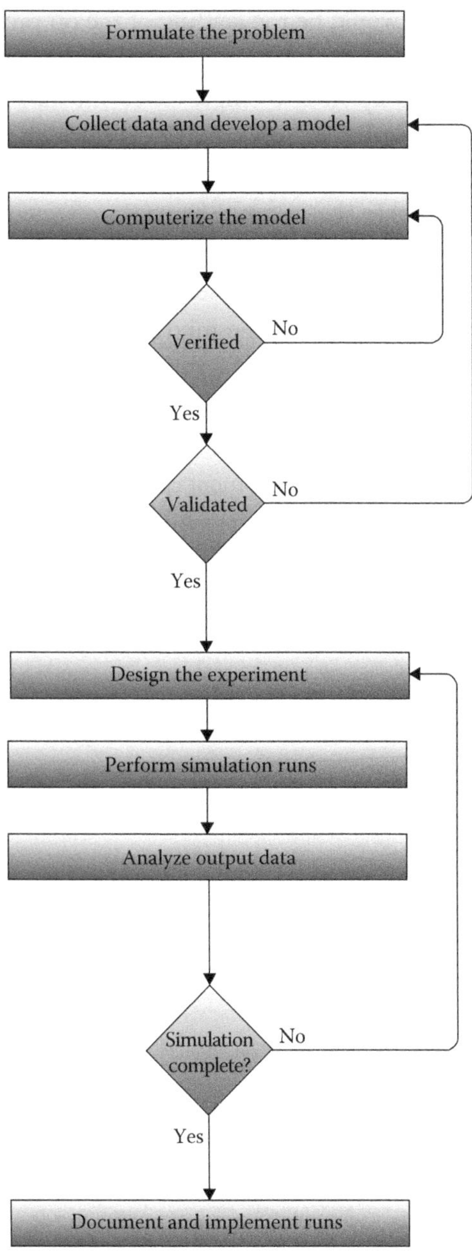

Figure 7.1 Simulation framework.

a match. They, then, contact a third-party inspector chosen by the customer to validate the match.

The following step would be the loading of the oil mixture in tanks, waiting for the arrival of the customer's vessel. The area known as *the industrial island* holds up to two ships. Once the vessel is ready, the filling process takes place. The mixture travels through three pipelines filling only 1 ft of the vessel, or about 10%. Now, a sample is taken to the lab to check whether the mixture has been affected. This effect is usually due to the vessel being corroded and, thus, requires that the customer cleans the vessel. Once the vessel is cleaned, the filling process would resume once more. Yet, if the mixture remains unaffected, the lab would send an approval to load the full vessel. When the loading is completed, another sample is taken before sealing the vessel. If the lab states that the mixture is altered, the process must start over again. One must note that this rarely occurs. In most cases, the lab tests are positive and a quality certificate, also known as bill of lading, is issued. When the customer receives the quality certificate, the products become ready for shipment; and the customer has a right to do whatever he or she wants with the product. This process is illustrated in Figure 7.2.

7.4 Arena Simulation

The refinery's simulated work flow can be seen in Figure 7.3.

7.4.1 Step-By-Step Process Identification

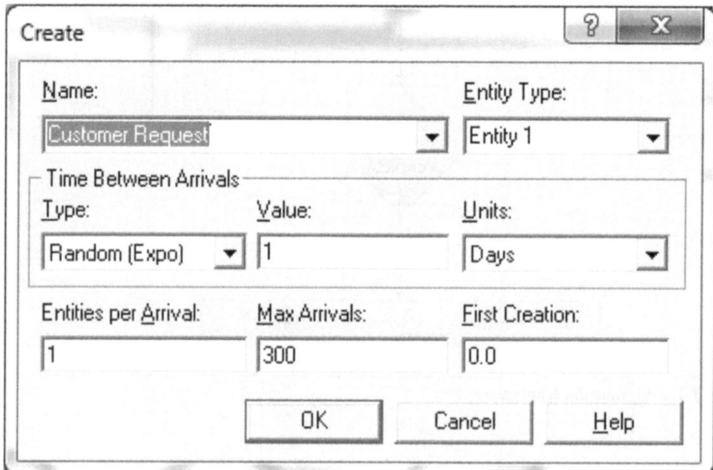
Create

Name: Customer Request
Entity Type: Entity 1

Time Between Arrivals
Type: Random (Expo)
Value: 1
Units: Days

Entities per Arrival: 1
Max Arrivals: 300
First Creation: 0.0

OK Cancel Help

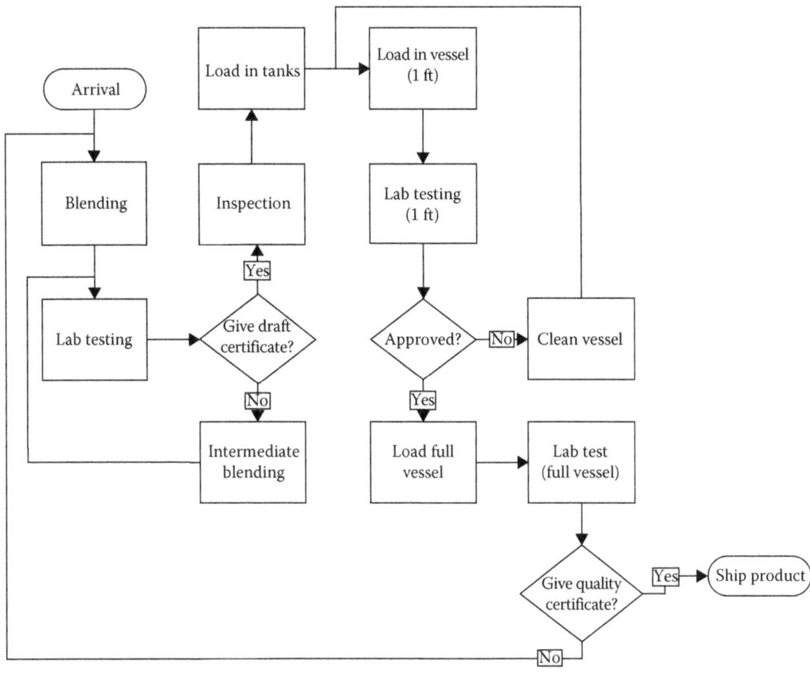

Figure 7.2 Refinery process flowchart.

The first process is the arrival of the customer request. The time to process this request is 1 day. The company stated that currently an average of 300 vessels is their output per year.

Hold

Name: Hold Request

Type: Scan for Condition

Condition: nq(Blending.Queue)<3

Queue Type: Queue

Queue Name: Hold Request.Queue

OK Cancel Help

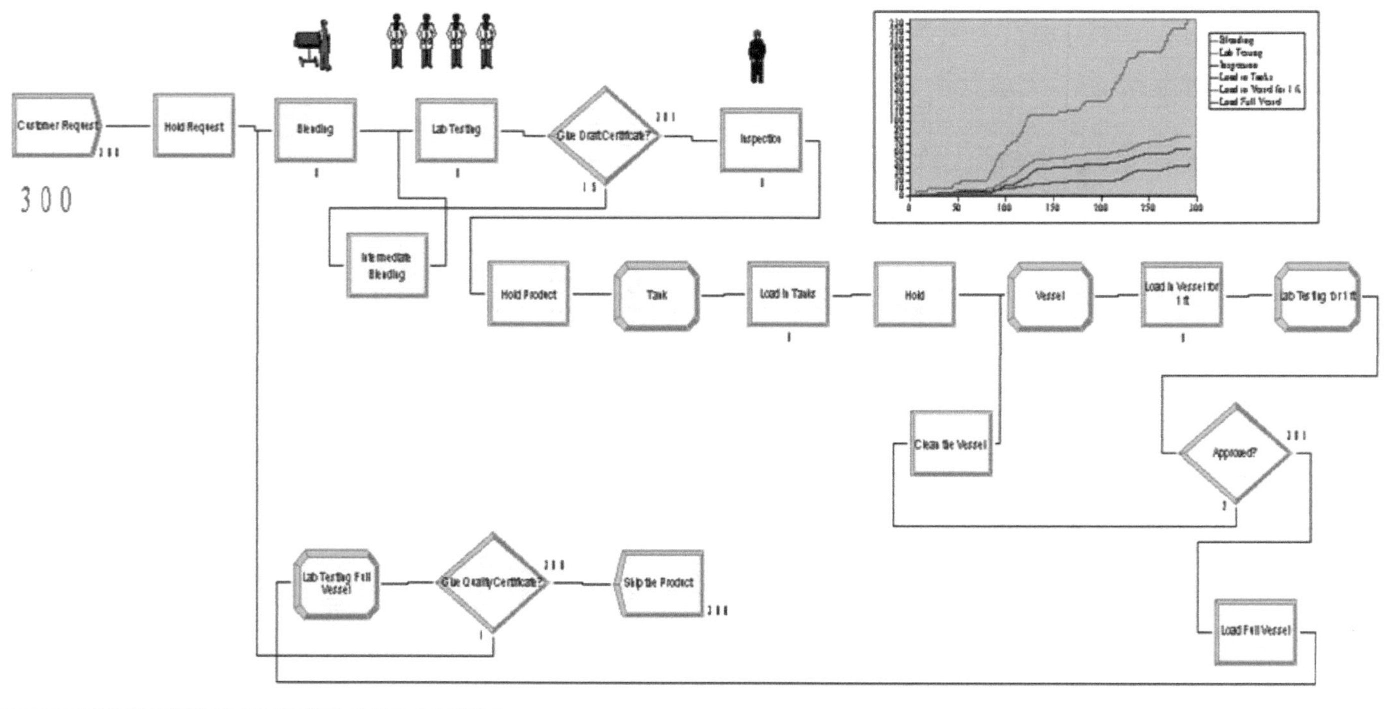

Figure 7.3 Arena simulation.

The company can process a maximum of three customer blends at a time; any blend coming while the three blends are being processed must be held in a queue.

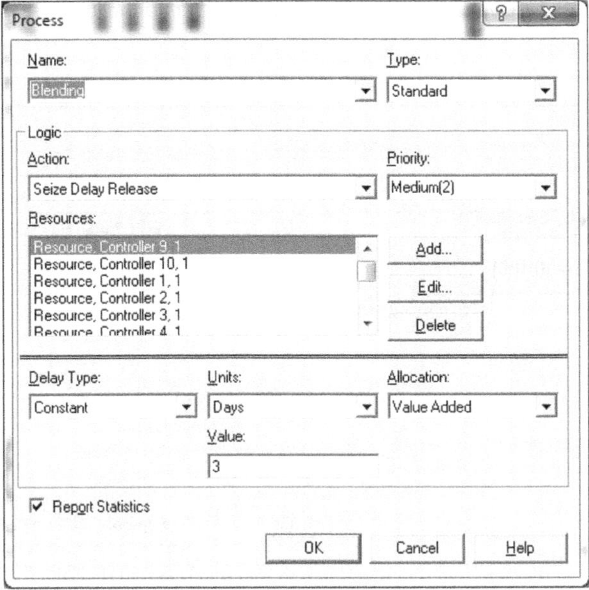

The blending process always takes 3 days to be completed. The process is performed by the controller unit, with the capacity of 4 employees in each 1 of the 10 units.

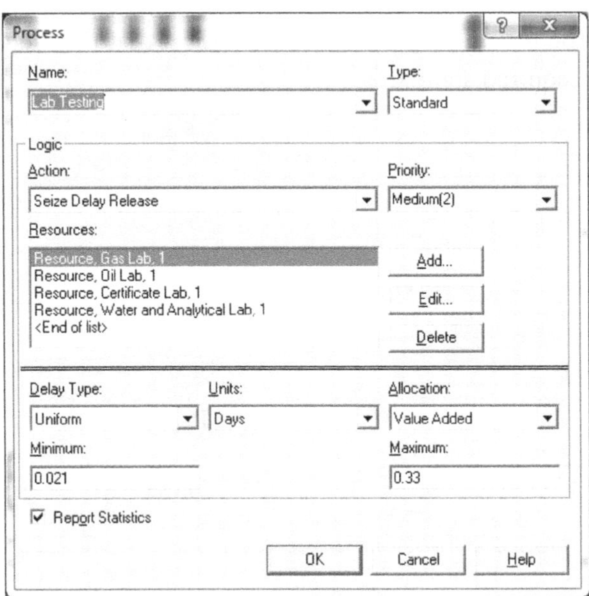

After the blending process, the employees need to take a spec to the lab for testing. There are four types of labs: gas lab, oil lab, certificate lab, and water and analytical lab. Each lab has 20 employees whose process takes between half an hour and 8 h to be completed.

Decide
Name:
Type:
Give Draft Certificate?
2-way by Chance
Percent True (0-100):
95
%
OK
Cancel
Help

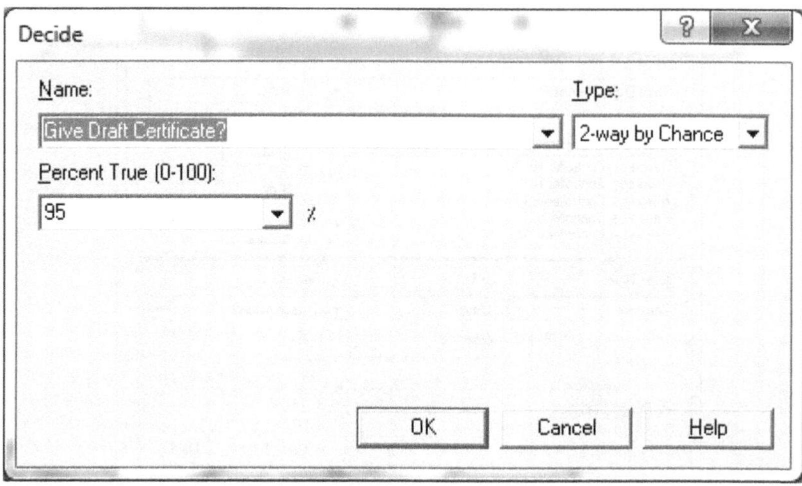

The decide process is a two-way process. There is a 95% chance of giving the spec a draft certificate and a 5% chance of rejecting the spec. The reason the lab rejects the spec is that the blend did not match the required standards.

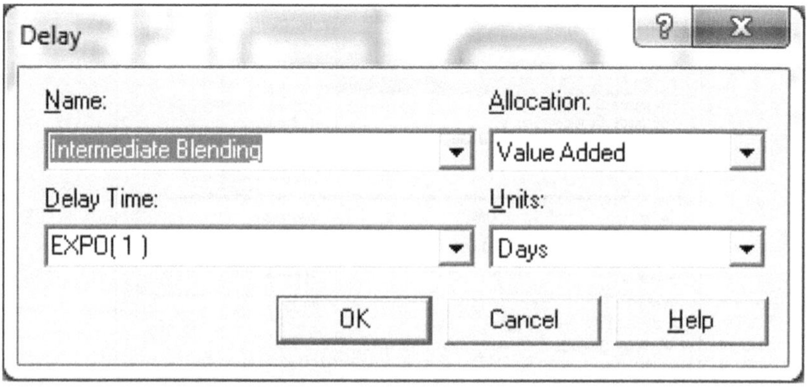

Delay
Name:
Allocation:
Intermediate Blending
Value Added
Delay Time:
Units:
EXPO(1)
Days
OK
Cancel
Help

When a spec is disapproved, it goes to the intermediate blending to change the blend to meet the specifications. This process is a delay process, because the blend takes additional time to be processed.

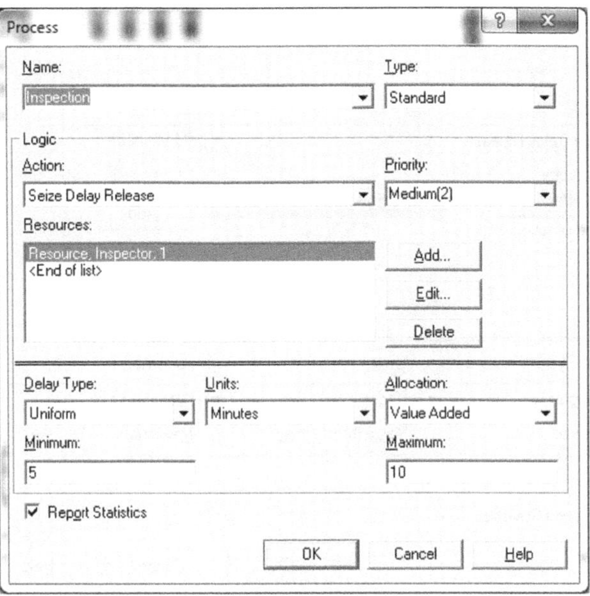

The third-party inspector comes to confirm that the spec matches the customer's requirements. He or she takes an average of 5–10 min to complete inspection (assuming all mixture must pass inspection).

The company has 50 holding/storage tanks.

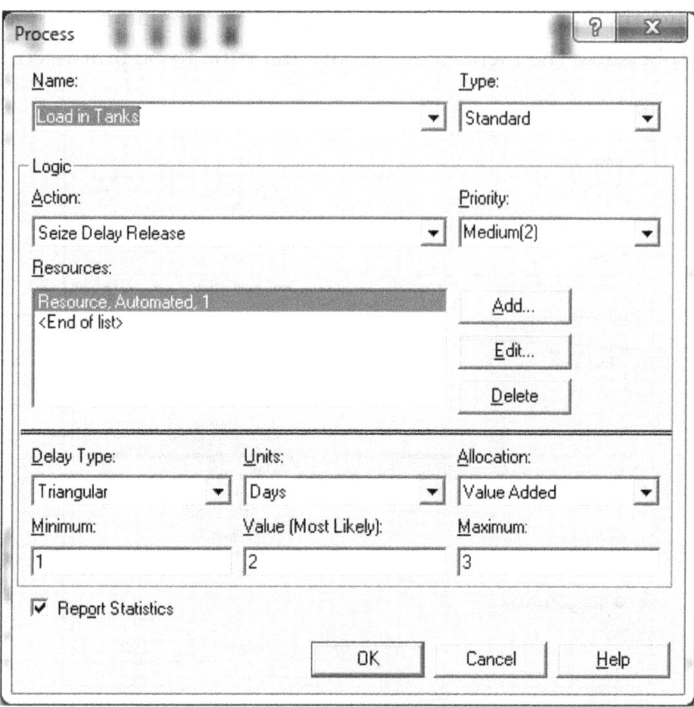

Loading in all tanks is an automated process that takes 1–3 days, most likely 2 days.

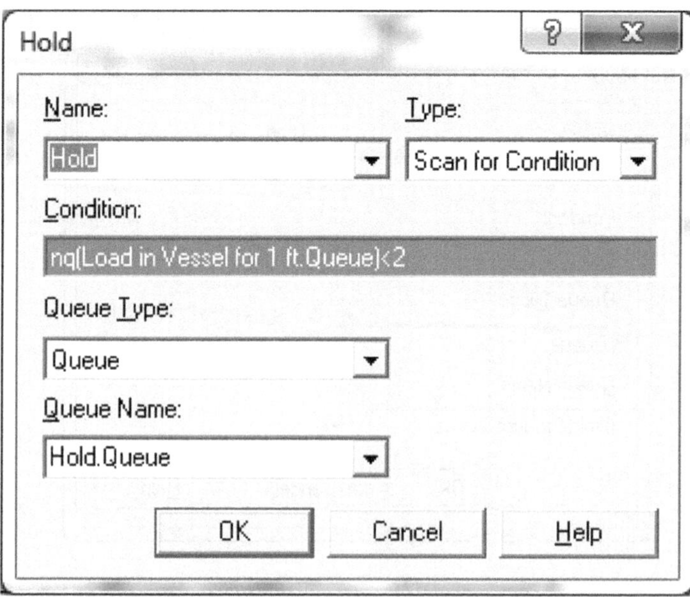

The industrial island, where the blends are loaded on to vessels, can hold two vessels at one time.

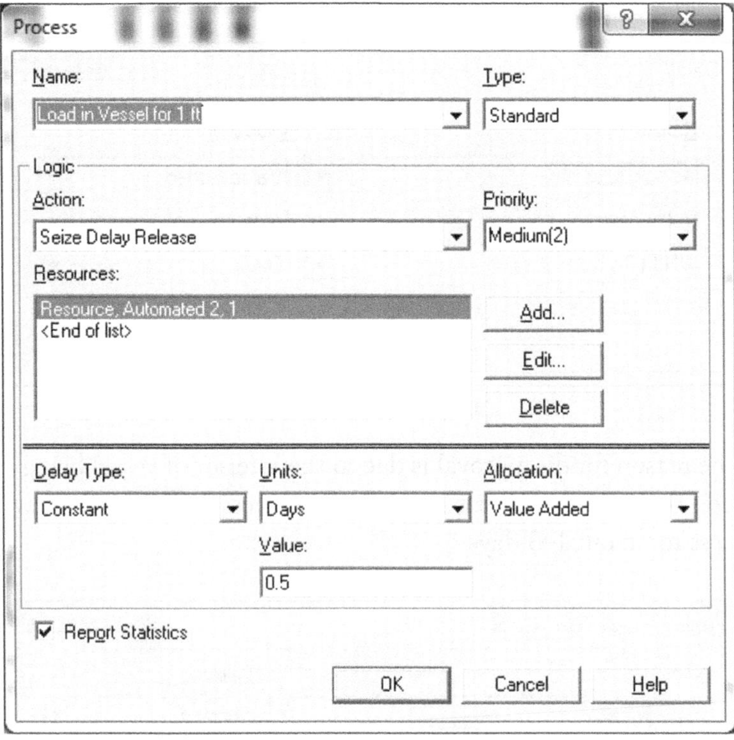

For loading the vessel, the company first has to load for 1 ft, roughly 10% of the vessel's volume. This process takes about half a day.

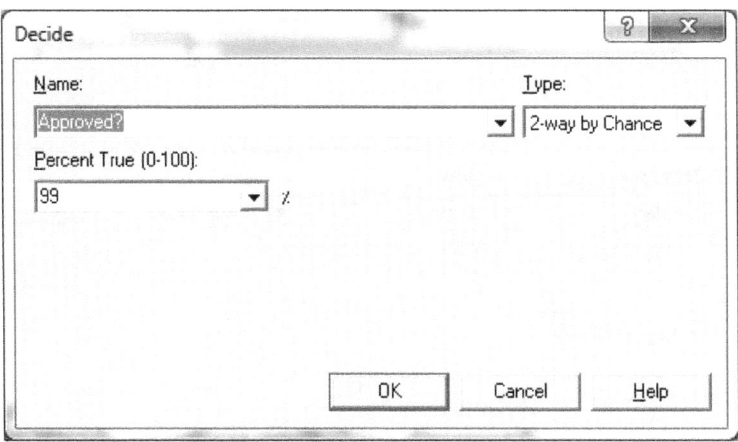

A lab test is, now, required to check if the blend has been altered. This shows that the vessel is suitable for full loading. Ninety-nine percent of the time, the spec is approved.

Delay
Name:
Clean the Vessel
Allocation:
Value Added
Delay Time:
UNIF(1 , 3)
Units:
Days
OK
Cancel
Help

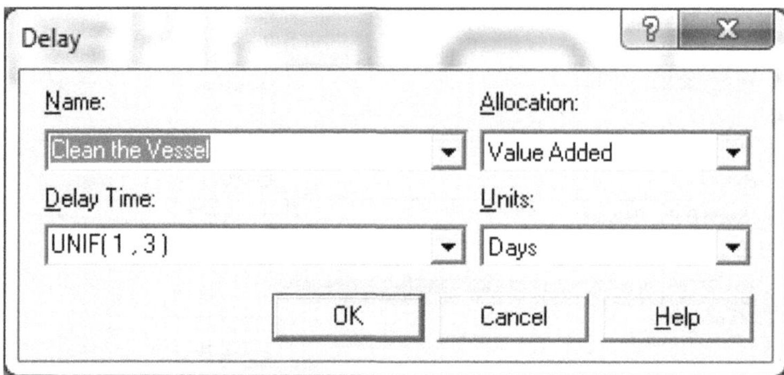

The reason for disapproval is due to the interior of vessel being corroded. The customer, then, must clean the vessel. This will delay the process for about 1–3 days.

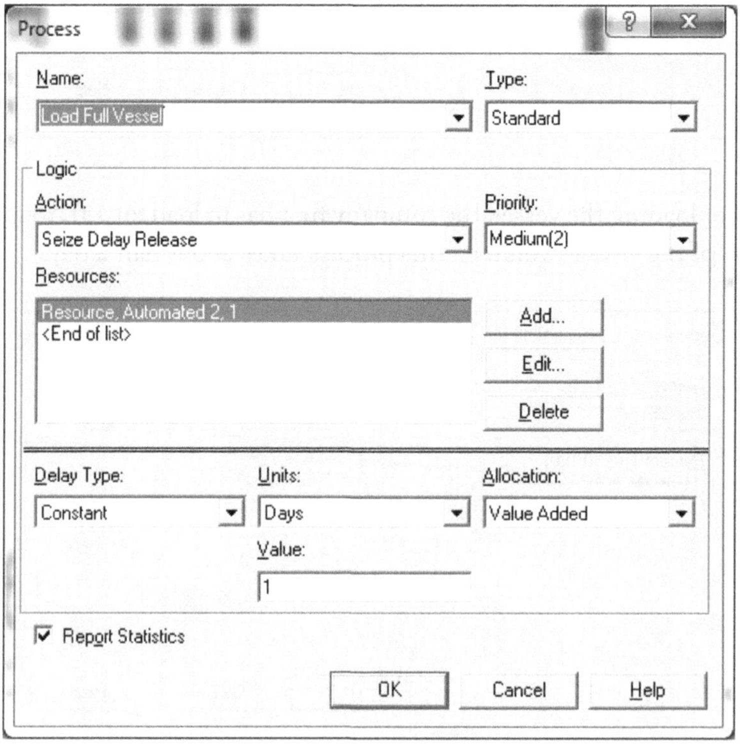

Process
Name:
Load Full Vessel
Type:
Standard
Logic
Action:
Seize Delay Release
Priority:
Medium(2)
Resources:
Resource, Automated 2, 1
<End of list>
Add...
Edit...
Delete
Delay Type:
Constant
Units:
Days
Allocation:
Value Added
Value:
1
Report Statistics
OK
Cancel
Help

If the approval is issued, the pipe will fill the vessel to its maximum capacity, taking 1 day to be completed.

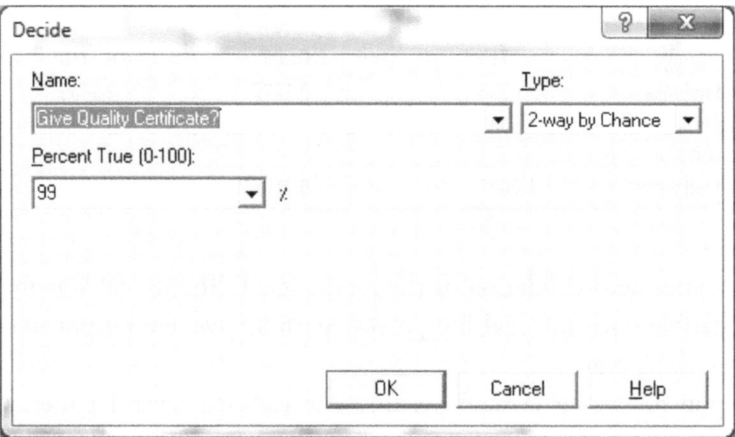

A final lab test will be done for the full vessel. One percent of the time, the quality certificate is not given. The product must be sent to blending to repeat the whole process once more. Arena counts the rework as an output.

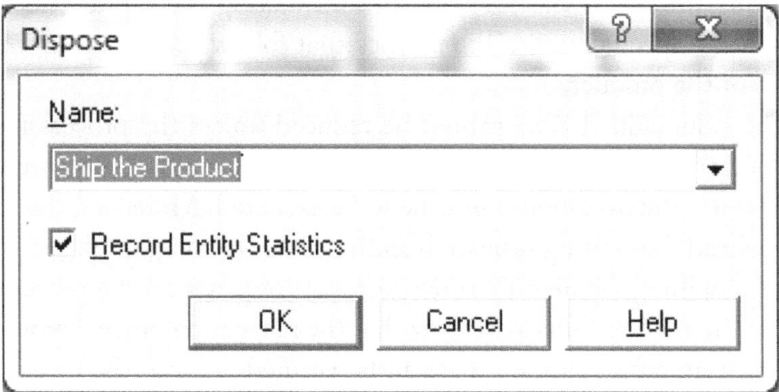

When a quality certificate is finally issued, the customer receives full responsibility of the product.

7.5 Output Analysis

Once one has simulated this complex production line, one is given results based on that simulation. The first step of analysis was the

Table 7.1 Processing Time in Each Process

ENTITY	VALUE-ADDED TIME (DAYS)	WAITING TIME (DAYS)	TOTAL TIME (DAYS)
Blending	3.0000	0.6313	3.6313
Inspection	0.00513163	0.00001087	0.00514250
Lab testing	0.1716	0.000	0.1716
Load in tanks	2.0038	0.1427	2.1465
Load in vessel for 1 ft	0.5000	0.2643	0.7643
Load full vessel	1.0001	0.2102	1.2102

verification and validation of the model itself. The model was found compatible with the real-life process with an average output of 300 vessels for 1 year.

Upon analyzing Arena's results, one can see several parameters being influenced by the process components such as time, capacity, and utilization.

In Table 7.1, one can note the total time spent in each process. From this, one can see how long the product remains in each entity and how long it waits to be processed. The process with the highest values in both is blending. This is accurate due to the fact that the blending process is the core step in formulating the product to suit customers' specifications. The entities that follow revolve around the blend of the product.

The value-added time cannot be reduced unless the process itself is changed. That is because within the petroleum industry, processes require a standard amount of time to be executed. Moreover, the reason behind the waiting time for blending is due to company possessing three pipelines, where only three customer requests can be processed at one time. The results also show that the inspection process requires the least amount of time. That can be verified because the inspector does not add to the product but simply checks if the blend specs match the customer's request. Figure 7.4 emphasizes the utilization of the company employees compared to the inspector's in relevance to the company's production line.

Loading 1 ft of the vessel has a value-added time of 0.500 days, while loading full vessel requires 1.0001 day. This can be justified because within the loading of the 1 ft process, an employee takes a sample from the vessel and sends it for lab testing.

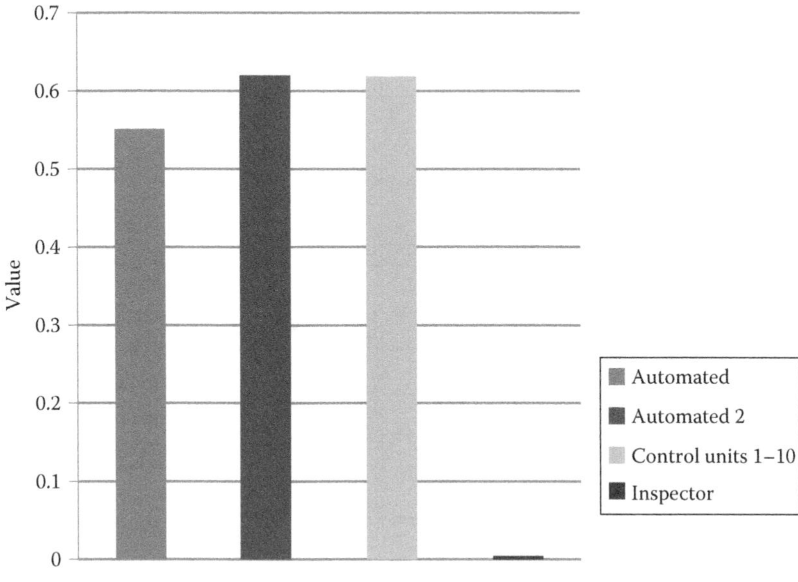

Figure 7.4 Utilization of resources.

7.5.1 Number In/Number Out

After simulating the process for 1 year, the numbers in and out are shown in Table 7.2. The average number of output is 300 vessels. However, the output of each process is greater due to necessary rework. For example, three blends were found to be altered after loading 1 ft of the vessel, considered by Arena as additional input.

Figure 7.5 emphasizes that lab testing has the highest number out with an output of 316 vessels. This means that an average of 16 samples was rejected during the simulated year. This is agreeable because it accounts for a tolerable 5.33% of total output.

Table 7.2 Number of In and Out after One Year

ENTITY	NUMBER IN/NUMBER OUT (VESSELS)
Average output	300
Blending	301
Inspection	301
Lab testing	316
Load in tanks	301
Load in vessel for 1 ft	303
Load full vessel	301

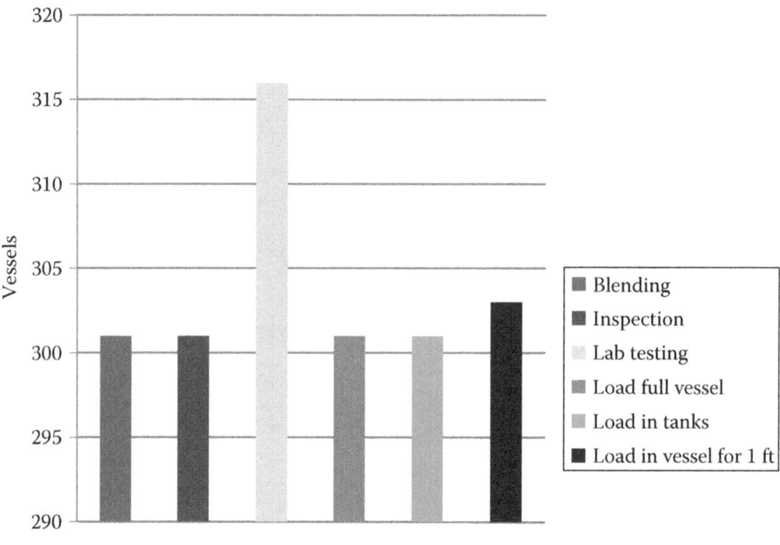

Figure 7.5 Number in/number out.

7.5.2 Parameter: Duration

The initial model focused on the duration of 1 year.* The company states that 300 vessels are their average output; this complies with the result of the simulated model. Though 300 vessels are the current average output, the simulations prove that the production line can withstand the maximum output number of 384 vessels. Hence, the company can further utilize their resources, if needed, to increase their production rate and therefore increase their profitability.

In reference to Table 7.3, one can see the output generated at a given amount of time. This is important for future knowledge in terms of profit. Given the number of vessels produced in a span of 10 years, the company can multiply the number of vessels by $X, which is the future price of an oil barrel. Thus, they can calculate their *expected* profit within 10 years. This can also be useful to know how much the company will lose if the refinery were to shut down for a week.

As of April 20, 2013, the price of an oil barrel (Kuwait News Agency, 2013) was $96.90. The company claims a vessel holds an

* The oil sector's year starts in April and ends in March.

Table 7.3 Generated Output

DURATION (DAYS)	NUMBER OUT
7	1 vessel
30	20 vessels
180	176 vessels
365	300 vessels
3650	3569 vessels

average of 20,000–60,000 barrels. Knowing this, one can deduce the following:

A refinery shut down for a week will result in an estimated loss of

$$\$96.90 \times 20{,}000 = \$1{,}938{,}000$$

$$\$96.90 \times 60{,}000 = \$5{,}814{,}000$$

The price of a vessel ranges from \$1,938,000 to \$5,814,000. The annual profit based on an output of 300 vessels ranges from \$581,400,000 to \$1,744,200,000.

7.5.3 Parameter: Employee Utilization

Human capital is a vital component within the oil sector. Employees formulate decisions and operate the production line. An employee in the control room has an average salary range of 2400–3000 KD. Ideally, a reduction in the number of employees would reduce labor cost. The control room in the refinery has 10 units, each with 4 employees.

Utilization of employees reflects on how efficiently a company utilizes its personnel. The allowed range of utilization is within 40% and 90%. As the percentage of utilization increases, this resource would be utilized more efficiently. Currently, the four employees of a unit have a utilization percentage of 61.85% as shown in Table 7.4. Though it is within the range, reducing each unit by one employee can save the company an average of 27,000 KD and increases the utilization of the employees to 83.01%, still within the acceptable range. A unit of two employees exceeds the acceptable limit reaching an unhealthy 99.95%.

Table 7.4 Utilization Percentage of Employees

NUMBER OF EMPLOYEES	UTILIZATION (%)	NUMBER BUSY
2	99.95	1.99
3	83.01	2.49
4	61.85	2.47
5	49.81	2.49
6	41.51	2.49

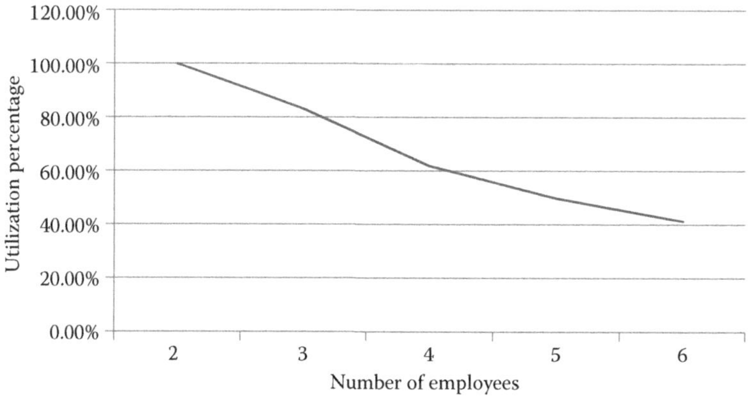

Figure 7.6 Employee utilization.

Though the idea of reduction in the number of employees is desirable, changing this parameter would increase the waiting time in the blending process from 0.6313 to 1.5578 days. Referring to Figure 7.6, one can notice that as the number of employees increases, the utilization percentage decreases.

7.5.4 Usage

Table 7.5 shows the utilization of workers and automated systems. The most utilized entity is Automated 2 with 61.99%. This is veritable since it controls two processes: loading vessel for 1 ft and loading full vessel. Automated follows with 55.08% since it fills 50 tanks. Regarding the 10 control units, each has a utilization of 61.85%, explained previously.

In reference to Figure 7.7, Automated 2 has an output value of 604 vessels. That is rational because it is in charge of loading 1 ft of the vessel as well as loading full vessels, summing up to 604. The remaining

Table 7.5 Utilization Percentage of Resources

RESOURCE	SCHEDULED UTILIZATION	TOTAL NUMBER SEIZED
Automated	0.5508	301
Automated 2	0.6199	604
Controllers 1–10	0.6185	301
Inspector	0.00423183	301

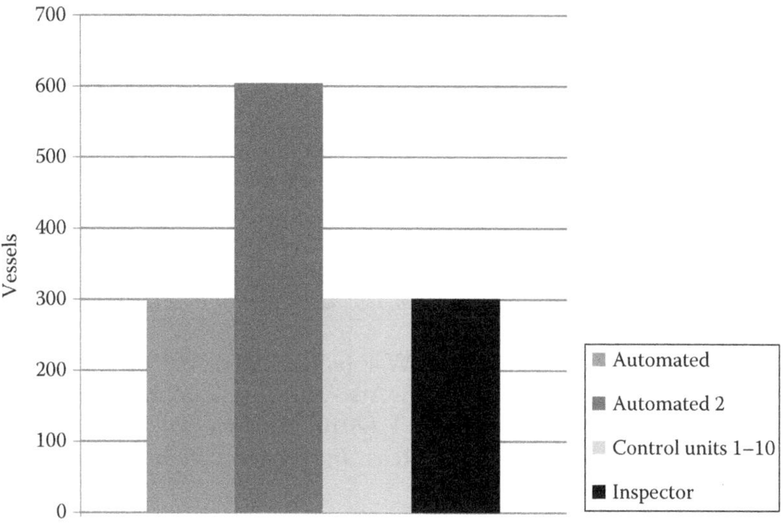

Figure 7.7 Total number seized.

resources match the number in/number out, indicating that the output of each resource matches the output of the company.

7.6 Summary and Concluding Remarks

A simulated model of a refinery's crude oil production line was provided to the company for future use. They can apply possible changes to analyze how they impact the real system instead of direct application.

The oil industry in Kuwait has complex production lines, integrated to provide Kuwait with 95% of its income. Each of Kuwait's petroleum companies is in charge of different operations within these production lines. The model, designed for the company, was verified and then validated with the actual system's average output of 300 vessels. Through simulation, results indicate that the refinery can manage a maximum average output capacity of 384 vessels.

Upon analyzing the statistical results of Arena, the current situation was vindicated. Every attribute was backed up with evidence that supports its case. Having the number of output per year provides knowledge of expected profit/loss. The simulation model will be contributed to the company to provide a basis for future modifications of their crude oil refining production line, where they can change processes without affecting the real system.

References

Carson, Y. and Maria, A. (1997). Simulation optimization: Methods and applications. *Proceedings of the 1997 Winter Simulation Conference*, Atlanta, GA, pp. 118–126.

Cortes, P., Onieva, L., and Guadix, J. (2009). Optimizing and simulating assembly line balancing problem in a motorcycle manufacturing company: A case study. *International Journal of Production Research*, 48(12), 2840–2860.

Groner, R., Groner, M., and Bischof, W. F. (eds.). (1983). *Methods of Heuristics*. Hillsdale, NJ: Lawrence Erlbaum Associates, Inc. Publishers, pp. 79–88.

Hecker, F., Hussein, W., and Becker, T. (2010). Analysis and optimization of bakery production line using arena. *International Journal of Simulation Modelling*, 9(4), 208–216.

Kleijnen, J. P. C. (2008). Computer simulation: Practice & research [PowerPoint slides]. http://www.tilburguniversity.edu/webwijs/files/center/kleijnen/TiLPS.pdf. Retrieved April 15, 2008.

Kuwait News Agency. (2013). Kuwaiti crude price rises to USD 96.9 pb. KUNA. http://www.kuna.net.kw/. Retrieved April 20, 2013.

Saad, S. M., Lau, K. H., and Omer, A. (2009). Design and analysis of oil production area—A simulation approach. *Proceedings of the 23rd European Conference on Modeling and Simulation*, Madrid, Spain, pp. 52–59.

8

Simulation and Analysis of Izmir's Metro Transportation System

ADALET ONER AND GULER OZTURK

Contents

8.1 Introduction

Izmir is the third largest city in Turkey with a population of approximately three million people. The city has a public transportation system that is a combination of land, sea, and underground networks. The underground transportation system (metro line) is 12 km in length and carries approximately 180,000 passengers a day on average. This chapter focuses on the analysis of operations in metro line via a simulation model. The analysis includes all the phases that a typical simulation study covers, such as data collection, input analysis, model building, verification and validation, experimental design, and output data analysis.

The goal is to build a simulation model to analyze and evaluate the effectiveness of operations in metro line. The performance criteria of effectiveness are the cost of energy spent for the operations (0.7 TL/coach/km), the average waiting time of passengers at stations, and the comfort of passengers (the number of straphangers in each square meter of the empty space in the coach).

8.2 Characteristics of Metro Line and Operations Management

The metro line has 10 stations. The first and last stations are terminuses, where trains begin and finish their trips in the line. The metro line lies in west–east direction; station #1 represents the west terminus, whereas station #10 sits at the east end of the line. The distance between stations varies between 0.8 and 1.6 km. The line has two tracks; therefore, trains can travel in both directions simultaneously.

Station #6 is the midpoint of this line, which consists of management center and maintenance facilities. Besides, it serves as a parking lot for trains. Trains access and leave the metro line at this station. Each train contains four coaches currently. There are 45 coaches in total, and each coach has 44 seats. There is additional 34 m^2 of empty space in each coach for standing passengers who cannot have a seat when the coach is crowded. Under congested situations, when all seats in a coach are taken, some more passengers can still get into the coach as straphangers are available in the empty space. The number of straphangers is limited to 6 per m^2. In such a case, each coach may have up to 248 passengers. This maximum capacity creates an uncomfortable voyage and is not desirable. The company wants to provide a service not to exceed 3 standing passengers per square meter of empty space, which means 146 passengers per coach.

The system begins operating at 6 o'clock every day, when trains start their daily tours. Traveling time between stations is known since the distance between stations is fixed. When a train arrives at a station, it waits for a specific period of time (dwell time) for passengers to get in and get off the train. Dwell times vary from station to station. There are 190 trips in each direction, which means 380 trips in total a day.

8.3 Literature Survey

There are many studies in the literature that involve transportation systems. Some studies are concerned with the analysis of the system whereas the others include the improvement and optimization of some performance criteria in the system.

Goverde (1998) dealt with the synchronization control of scheduled train services to minimize passenger waiting times. This model can be utilized to review the optimal synchronization control policy from the point of view of cost of arrival delays. The objectives are to minimize the total generalized passenger waiting time and to resolve buffer times related to the passenger waiting time in service network timetables.

Chang et al. (2000) are concerned with a multiobjective model for passenger train services planning that is applied to Taiwan's high-speed rail line. The goal is to improve a multiobjective programming model for the optimal assignment of passenger train services on an intercity high-speed rail line without branches.

Li (2000) built up a simulation model of a train station and its passenger flow. The simulation model consists of the processes, equipment, and queues a passenger comes across from entering the station. All these parameters directly affect the total passenger travel time. Minimizing the total passenger travel time and increasing the service quality are the purpose of this chapter.

Martinez (2002) proposed the application of Siman ARENA™, which is a discrete event simulation tool in the operational planning of a rail system. This model also involves an animation of a Siman simulation. The simulation model gives the capability of using a realistic model of a rail network and calculated waiting time in the platform and on-time performance for special system performance parameters.

Sheu and Lin (2011) presented the optimization of train regulation and energy usage of metro lines using an adaptive—optimal—control algorithm. The automatic train regulation system involves service quality, transport capacity, and energy usage of metro line operations. The train regulator has a purpose of maximizing the schedule and headway commitment while minimizing the energy depletion.

Finally, Yalçınkaya and Bayhan (2009) described the modeling and optimization of average travel time for a metro line by simulation. They present a modeling and solution approach based on simulation and response surface methodology for optimization. The aim is to find the optimum headways and to minimize the average passenger time spent in the metro line with a satisfactory rate of carriage fullness. Actually, their study is conducted on the same transportation system as in our model. There are some discrepancies in that study such that the arrival rates of passengers are assumed to be fixed during the day. However, it is shown that the arrival rates are nonstationary. On the other hand, the objective is to find the optimal headways (time interval between consecutive trips at a station). They assume that there are an infinite number of trains available at the end of the metro lines. In fact, there are a limited number of trains available and the headways are limited by some other factors.

8.4 Input Analysis

There are two stochastic inputs of the system. The first one is the arrival of passengers into the system, and the other one is the destination station of the passengers. The details of input analysis related to each input are given as follows.

When a passenger arrives at a station, he/she uses an electronic pass card or a token to access the system. Therefore, the arrival information, including the time stamp and station, is automatically stored in a database. The records of passengers served by the system within last 15 months have been considered for the analysis in order to determine arrival rates. There are more than 50 millions of such records in the database. The average number of passengers has been determined for every minute of a day, for each day of the week, and for each station. The idea is to identify the differences, if any, in arrival rates for different stations and hours in a day. We are also interested to know whether there is a difference between days of a week (working days vs weekend) and between winter and summer seasons.

It has been observed that every station has its own characteristic pattern for arrival rates. It is not surprising since the stations are located in different parts of the city. Some stations are near the business center whereas some others are in residential areas. The arrival patterns for two different stations are given in Figure 8.1, which depicts the average number of passengers per minute that arrive at those stations over time during the day.

Another remarkable observation is that the patterns repeat themselves at each particular station during weekdays but differ only

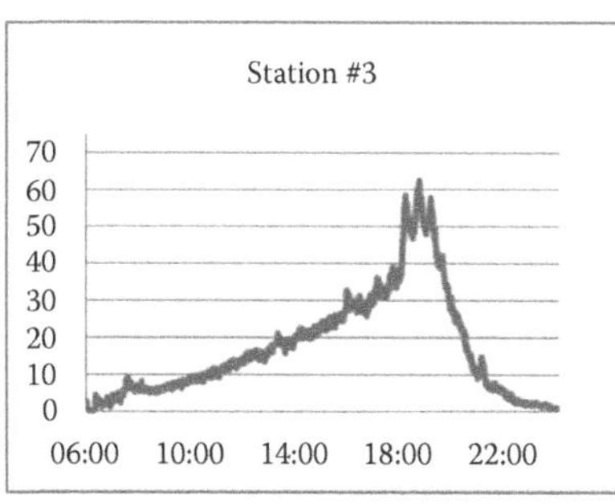

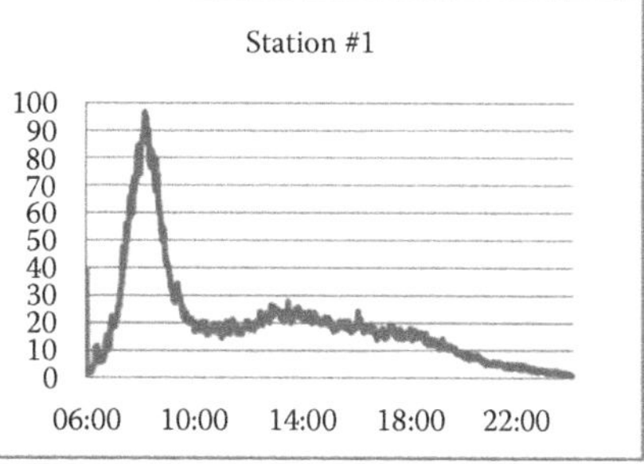

Figure 8.1 The average number of passengers per minute at two different stations on Monday.

during weekends, Saturday and Sunday. It leads to an important result such that there should be three scenarios in the model: one for weekdays, the one for Saturdays, and the last one for Sundays. The final observation is that only the magnitude changes, whereas the shapes of the patterns do not change when seasonal changes come into question.

Another stochastic input is the destination of passengers. When a passenger accesses the system at a particular station, there is no information available about at which station he/she leaves the metro line. However, it is important to have this information in order to assess the performance of the system.

One-week lasting public surveys have taken place twice during winter and summer seasons in order to provide information about the destination stations of the passengers. The surveys are conducted with the assistance of the security personnel. At each station, security men questioned the arriving passengers about their destination station. The answers, including time stamps, have been automatically recorded with their hand counters. Each survey covers more than 500,000 passengers in total. The collected data have been organized and analyzed to produce probability tables for destination stations. The outcomes are assumed to be confident and consistent since sample sizes are large for each station.

Destination probabilities are organized in two complementary tables for each station, because when a passenger arrives at a station, it is required to identify the probability of the direction that he/she will take. The probability of destination station should be determined next. If the station is a terminus, it is obvious that the passenger will travel in either west or east direction depending on the terminus he/she arrives. However, if the station is an intermediate one, a preliminary probability should be assigned for the direction, and then a probability table should accompany for the destination stations.

Due to socioeconomical behavior of the passengers and the attributes of the stations in the line, destination probabilities may change over time during the day. Therefore, the tables should be prepared as a function of time to represent those fluctuations.

8.5 Simulation Model

Simulation model is created by ARENA software, which uses discrete-event simulation methodology. An animation model is also established to accompany the simulation model. The model is accompanied by two external files, one for inputs and the other for outputs of the simulation. Both of them are MS Excel™ files. The input file includes travel durations for trains between stations, dwell times at stations, current train schedule, arrival rates at each station at each minute, the probabilities of direction and destination stations of passengers arriving at each station as function of time, and the number of trains in the system over time during the day.

The model is developed in a modular structure that includes several submodels. For example, a submodel is concerned with the arrival of passengers, that is, creating passenger entities. Statistical input analysis reveals that the arrival rates are constantly changing during a day. It means that it is not appropriate to investigate a single probability distribution to fit for interarrival times. It can be shown that arrivals should be modeled using nonstationary Poisson process. Let $\lambda(t)$ be the arrival rate of passengers to a particular station at time t. Parameter $\lambda(t)$ is not constant but changes over time. For example, it would be larger during the morning rush hour than in the middle of the afternoon. In our model, the input analysis of arrivals has been summarized in a table that consists of the average number of arrivals per minute for each station. The entries of that tables correspond to parameter $\lambda(t)$ for nonstationary Poisson process at minute t. Since the passengers are created using *Poisson process*, the time between arrivals should come from the exponential distribution with mean $(1/\lambda(t))$. The simulation model reads the average number of passengers from the input file and then assigns to corresponding parameter $\lambda(t)$ for that particular minute. For the following minutes, different values are read from the input file, which represent different values for parameter $\lambda(t)$. Those values are then assigned to corresponding variables in the simulation model.

When a passenger arrives at an intermediate station, first of all, it is required to identify the probability of the direction he/she will take. Then the probability of destination station should be known. Furthermore, those probabilities may change over time during the day; therefore, they have to be identified as a function of time. All those probability tables are predetermined through input analysis and stored in the input file. The simulation model reads those values and assigns them to corresponding variables.

There are other submodels that take care of other activities in the system. One of them simulates train motions and their entrance into and exit from the system. It also controls time, distance between stations, and the speed of the trains. Another submodel controls special statistical counters and system parameters. Finally, a submodel that controls the interaction between trains and passengers has been developed for each station.

8.6 Verification and Validation

In simulation studies, the concept of *verification* is concerned with building the *model right*. It is used in the comparison of the conceptual model with the computer representation. It asks the following questions: "Is the model implemented correctly in the computer? Are the input parameters and logical structure of the model correctly represented?"

In order to verify the model, each module and unit has been tested using the debugging tools of ARENA software during the model development process. The movements of particular passengers are tracked upon arrival at the system. Choosing direction and identifying destination station and then putting the passenger into the appropriate queue has been verified by tracking random passengers at each station. Furthermore, especially the train–passenger interactions are investigated carefully. Technically, trains and passengers are two different kinds of entities in ARENA. When a train is ready to move from a terminus, the entities representing the passengers getting in the train and the entity representing the train are merged into a group and they are moved together to the next station. Upon arriving at the next station, some of the entities should be separated from the group to represent the leaving passengers. Some new entities are then merged with the group to simulate the passengers getting in the train. The movements of trains are tracked over time as to whether they comply with regulations and time schedule.

On the other hand *validation* is concerned with building the *right model*. It is used to determine that a model is an accurate representation of the real system. Validation is usually achieved through the calibration of the model, an iterative process of comparing the model with the actual system behavior and using the discrepancies between the two, and the insights gained, to improve the model. This process is repeated until the accuracy of the model is judged to be acceptable.

Some comparisons have been prepared in order to validate the model. Actual and simulated numbers of passengers arriving at the stations are compared, and it is found that they are very close to each other. Table 8.1 shows the comparison related to the total passengers at each station.

Furthermore, the numbers of arriving passengers are also compared on an hourly basis at each station. Table 8.2 presents the comparisons for two selected stations at different hours of the day.

Table 8.1 Comparison of the Total Number of Passengers Arriving at Stations

STATION NAME	SIMULATION AVERAGE	ACTUAL SYSTEM AVERAGE
Station #10	29,943	30,028
Station #9	7,296	7,321
Station #8	2,706	2,722
Station #7	10,048	10,059
Station #6	11,460	11,503
Station #5	1,118	1,132
Station #4	7,904	7,930
Station #3	18,735	18,780
Station #2	25,446	25,516
Station #1	21,257	21,267
Total	135,913	136,258

Table 8.2 Comparison of the Hourly Number of Passengers Arriving at Stations

	STATION #1			STATION #2		
HOURS	GENERATED BY SIMULATION	ACTUAL	DEVIATION (%)	GENERATED BY SIMULATION	ACTUAL	DEVIATION (%)
05:30–07:00	547	531	3	393	394	0
07:00–08:00	2985	2920	2	1571	1547	2
08:00–09:00	4390	4419	–1	2094	2109	–1
09:00–10:00	1688	1712	–1	1260	1271	–1
10:00–11:00	1085	1083	0	1000	998	0
13:00–14:00	1344	1345	0	1605	1602	0
14:00–15:00	1252	1252	0	1687	1684	0
16:00–17:00	1048	1055	–1	2322	2315	0
17:00–18:00	997	1000	0	2567	2573	0
18:00–19:00	890	887	0	2299	2304	0
19:00–20:00	642	645	0	1762	1781	–1
22:00–23:00	202	202	0	442	445	–1
23:00–00:00	117	121	–3	255	265	–4

Please note that actual and simulated numbers of passengers arriving at the stations are very close to each other, which enable us to be confident with the verification and validation of the model. Furthermore, the animation model also provides a visual tool for validating the model. It is especially very helpful to present the model to the management of the company.

8.7 Output Analysis

The model developed in this chapter is *terminating simulation* since metro operations are stopped at midnight each day and start with an empty state the next day. It is required to collect data across some replications and analyze them statistically to construct confidence intervals for performance measures in the model. The performance measures in consideration are the number of passengers in the train across stations for each trip and the average waiting times of passengers at stations.

The number of replications is determined to be 10 in the beginning as proposed by Law and Kelton (2000). Table 8.3 indicates the confidence intervals for the number of passengers in the train across stations for selected trips.

Table 8.3 Confidence Intervals for the Number of Passengers in Train across Stations (Number of Replications = 10)

TRIP TIME	STATION #10	STATION #9	STATION #8	STATION #7	STATION #6	STATION #5	STATION #4	STATION #3	STATION #2
06:00	21.8 ± 3.61	26.1 ± 3.18	27.8 ± 2.99	33 ± 4.19	26.9 ± 3.36	26 ± 3.16	23.5 ± 3.02	19.1 ± 3.34	9.5 ± 2.55
06:10	10.8 ± 2.33	14.1 ± 1.95	14.4 ± 2.13	16.9 ± 3.06	13.4 ± 2.50	13.5 ± 2.26	16.9 ± 2.39	14.9 ± 2.59	11.6 ± 2.78
...	...	...	...	...	...	...	...	...	...
08:06	184.8 ± 7.39	232 ± 9.53	237 ± 10.8	278.9 ± 8.96	281.4 ± 8.04	274.1 ± 8.35	252.3 ± 10.3	175.6 ± 9.13	80.8 ± 3.67
08:10	205.4 ± 10.1	254.5 ± 10.5	255.6 ± 9.13	303.9 ± 12.5	311.2 ± 15.1	304.3 ± 14.2	279.5 ± 11.6	187.8 ± 5.64	80.9 ± 3.06
...	...	...	...	...	...	...	...	...	...
13:18	220.2 ± 10.1	246.8 ± 12.0	251.5 ± 12.4	299.3 ± 13.2	273.1 ± 14.3	270.4 ± 14.4	251.8 ± 14.6	222.2 ± 10.2	106 ± 8.98
13:24	195.9 ± 10.9	220.1 ± 9.64	229.3 ± 10.4	269.2 ± 10.9	269.5 ± 12.6	266.5 ± 12.9	253.2 ± 13.1	221.8 ± 14.0	113 ± 7.22
...	...	...	...	...	...	...	...	...	...
17:42	161.3 ± 11.6	184.9 ± 13.0	192.3 ± 12.5	215.7 ± 13.9	203.3 ± 13.1	201.3 ± 12.7	200 ± 12.0	214.2 ± 11.7	137 ± 4.53
...	...	...	...	...	...	...	...	...	...
20:47	50 ± 5.48	55.5 ± 6.03	58.3 ± 7.66	62.7 ± 6.79	73 ± 0.01	73 ± 0.02	71 ± 0.01	88 ± 0.03	65 ± 0.04
...	...	...	...	...	...	...	...	...	...

The averages and half-widths of confidence intervals differ substantially with respect to the trip time. In order to compare the significance of the confidence intervals in a common ground, the estimates of *relative error* are calculated for each trip. The estimate of relative error is defined as the division of half-width by the average (Law and Kelton, 2000). Table 8.4 shows the estimates of relative errors for the same trips given in Table 8.3.

Estimates for relative errors are very small for all trips except the ones that are the early trips in the morning and the late trips in the evening. It is not surprising since the number of passengers arriving at those hours is relatively small for every station, which leads to smaller averages. Confidence intervals and estimates of relative errors are also determined for the average waiting times of passengers at different stations.

The estimates of relative errors derived from 10 replications are reviewed for all system parameters, and it is observed that the highest relative error occurs for the number of passengers in the train at trip with time 06:00 am in west direction (see Table 8.4). This parameter will be used as the reference point to find the number of replications required to provide a smaller relative error level, which is determined to be 0.05. In other words, we need to figure out the number of replications to decrease the half-widths such that relative error is less than 0.05 for each system parameter.

The concept of relative error is defined by Law and Kelton (2000). If the estimate $X(n)$ is such that $|X(n)-\mu|/|\mu| = \gamma$, then we say that $X(n)$ is a relative error of γ. Suppose that we make replications of a simulation until the half-width of the confidence interval divided by $|X(n)|$ is less than or equal to γ $(0 < \gamma < 1)$. This ratio is an estimate of the actual relative error. It is shown that an approximate expression for the number of replications $n_r(\gamma)$ required to obtain a relative error of γ is given by

$$n_r(\gamma) = \text{Min}\left\{ i \geq n : \frac{t_{i-1,1-(\alpha/2)} * s(n)/\sqrt{i}}{X(n)} \leq \gamma' \right\}$$

where

$\gamma' = \gamma/(1+\gamma)$ is the adjusted relative error

$n_r(\gamma)$ is approximated as the smallest integer i satisfying $i \geq s^2(n) [z_{1-(\alpha/2)}/\gamma' * X(n)]^2$

Table 8.4 Estimates of Relative Errors (Half-Width Divided by Average)

TRIP TIME	STATION #10	STATION #9	STATION #8	STATION #7	STATION #6	STATION #5	STATION #4	STATION #3	STATION #2
06:00	0.17	0.12	0.11	0.13	0.13	0.12	0.13	0.18	0.27
06:10	0.22	0.14	0.15	0.18	0.19	0.17	0.14	0.17	0.24
…	…	…	…	…	…	…	…	…	…
08:06	0.04	0.04	0.05	0.03	0.03	0.03	0.04	0.05	0.05
08:10	0.05	0.04	0.04	0.04	0.05	0.05	0.04	0.03	0.04
…	…	…	…	…	…	…	…	…	…
13:18	0.05	0.05	0.05	0.04	0.05	0.05	0.06	0.05	0.08
13:24	0.06	0.04	0.05	0.04	0.05	0.05	0.05	0.06	0.06
…	…	…	…	…	…	…	…	…	…
17:42	0.07	0.07	0.07	0.06	0.06	0.06	0.06	0.05	0.03
…	…	…	…	…	…	…	…	…	…
20:47	0.11	0.11	0.13	0.11	0.0001	0.0002	0.0001	0.0003	0.0006
…	…	…	…	…	…	…	…	…	…

If we use the outcomes of our model, the number of replications to be done is found to be $n_r(\gamma) = 168$. On the other hand, there are some useful global functions that can handle confidence intervals and half-widths across replications in ARENA software. The built-in function *ORUNHALF* (*parameter*) delivers half-width of a system parameter after n replications. Furthermore, it is possible to develop an experimental design in ARENA such that the model itself makes replications sequentially until a desired level of half-width is attained and then it stops making further replications. Unfortunately, there is not such a function to work with relative errors. However, it is possible to translate the desired level of relative error into desired level of half-width as follows:

$$\frac{\text{Half-width}}{|X(n)|} \leq \frac{\gamma}{1+\gamma} = \gamma' \rightarrow \text{Half-width} \leq \gamma' * |X(n)|$$

The model was run with the preceding construct, and it stopped after 217 replications. The number of replications here is very close to the one found earlier using the procedures described by Law and Kelton (2000). The relative errors have been calculated then, and it is observed that the confidence intervals and half-widths are all much narrower than the previous ones, and the relative errors are all less than or equal to 0.05. The results derived after 217 replications are used as the formal outcomes of the simulation model.

8.8 Results and Discussions

The primary outcome of the simulation model is the average number of passengers in the train across the stations for each trip. It serves an indication for the comfort of the passengers. Figure 8.2 presents the average number of passengers in the train across stations for a particular trip.

The numbers in the figures represent the number of passengers in the train whenever the train leaves a station. Please remember that each train has 4 coaches and each coach has 44 seats plus 34 m^2 empty spaces for standing passengers. Note that the maximum numbers of passengers in the trains are 270 and 304 for the east and west rounds,

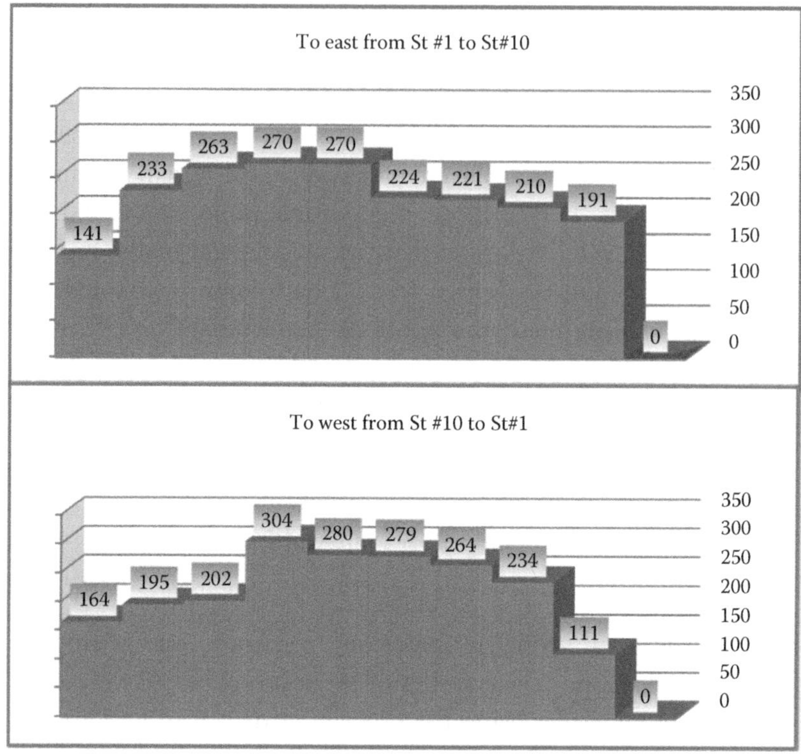

Figure 8.2 The average number of passengers in a train across stations for trip at 13:00.

respectively. In order to define the service quality clearly, let us make some definitions before proceeding further:

$CoI_{i\cdot d}$: Comfort index. The number of straphangers per square meter for trip i to direction d. $i = 1, 2, \ldots, 190$; $d =$ East, West

$MNP_{i\cdot d}$: Maximum number of passengers in the train for trip i

NS: Number of seats per coach

$NC_{i\cdot d}$: Number of coaches

ES: Empty space for straphangers per coach in terms of square meter

$$CoI_{i\cdot d} = \frac{MNP_{i\cdot d} - (NC_{i\cdot d} * NS)}{(NC_{i\cdot d} * ES)}$$

The service quality is defined in terms of comfort index. Quality gets worse as comfort index increases. The threshold level is determined to be 3. In other words, it is desired that comfort index should not exceed 3, and the number of coaches should be determined accordingly. The comfort indexes of the trip at 13:00 calculated for different number of coaches are as follows:

$$\mathrm{CoI}_{13:00\text{-west}} = \frac{304-(4*44)}{(4*34)} = 0.94 \quad \text{for 4 coaches}$$

$$\mathrm{CoI}_{13:00\text{-west}} = \frac{304-(3*44)}{(3*34)} = 1.68 \quad \text{for 3 coaches}$$

$$\mathrm{CoI}_{13:00\text{-west}} = \frac{304-(2*44)}{(2*34)} = 3.17 \quad \text{for 2 coaches}$$

The minimum number of coaches within comfort limits is determined to be three coaches, which means three coaches may be used for this trip instead of four coaches. Therefore, there is an opportunity to save one coach for this trip, which might decrease the energy cost by 8.4 TL (=12 km × 1 coach × 0.7 TL).

Comfort indexes are calculated for each trip in each direction, and the minimum number of coaches whose comfort index is less than 3 is determined. The results are given in Table 8.5.

The daily cost of energy for the current mode of operations that use four coaches for every trip can be calculated as follows:

$$\text{Cost(current)} = (0.7\ \text{TL/km/coach}) * (380\ \text{trips}) * (4\ \text{coaches/trip}) * (\approx 12\ \text{km/trip}) \approx 12{,}750\ \text{TL}$$

On the other hand, if the suggested numbers of coaches are used for the corresponding trips, then the daily cost of energy is as follows:

$$\text{Cost(suggested)} = (0.7\ \text{TL/km/coach}) * [(448+408)\text{coaches}] * (\approx 12\ \text{km/trip}) \approx 7190\ \text{TL}$$

It is clear that daily saving in the cost of energy is 5,560 TL and yearly saving could reach up to 1,389,000 TL. It implies remarkable improvement in the efficiency of the metro operations.

Another performance measure is the average waiting times of passengers at stations. Confidence intervals have been constructed, and

Table 8.5 Current and the Suggested Number of Coaches

TRIP NO.	TRIP TIME	MAXIMUM NUMBER OF PASSENGERS IN TRAIN	CURRENT NUMBER OF COACHES	CURRENT COMFORT INDEX	SUGGESTED NUMBER OF COACHES	SUGGESTED COMFORT INDEX
1	06:00:00	41	4	0	2	0.00
2	06:10:00	27	4	0	2	0.00
3	06:20:00	71	4	0	2	0.00
4	06:30:00	150	4	0	2	0.91
5	06:40:00	162	4	0	2	1.09
6	06:50:00	187	4	0.00	2	1.46
7	07:00:00	284	4	0.79	2	2.88
8	07:05:00	157	4	0.00	2	1.01
9	07:10:00	176	4	0.00	2	1.29
...	...	...	...	...	...	...
11	07:20:00	241	4	0.47	2	2.24
13	07:30:00	328	4	1.11	3	1.92
14	07:34:00	288	4	0.83	2	2.94
15	07:38:00	322	4	1.07	3	1.86
16	07:42:00	341	4	1.21	3	2.04
17	07:46:00	341	4	1.21	3	2.05
18	07:50:00	357	4	1.33	3	2.21
19	07:54:00	384	4	1.53	3	2.47
20	07:58:00	405	4	1.68	3	2.68
...	...	...	...	...	...	...
Current number of coaches used per day			380 × 4 = 1520	Suggested number of coaches used per day	448 + 408 = 856	

relative errors have been calculated for this measure. Table 8.6 summarizes the outcomes of the average waiting times.

Notice that all the confidence intervals and half-widths are small. The highest relative error in table is 0.018 for 217 replications. It is by far smaller than 0.05. It is observed that the highest value is 232 s ≈ 4 min for station #10.

Another interesting observation is the transfer times of the passengers, which indicate the average travel times for the passengers by the stations. The outcomes are given in Table 8.7.

The details of the model, outcomes, and recommendations are presented to the management of the Metro Company. Although the study is appreciated, there are discussions on the recommendations about

Table 8.6 Confidence Intervals and Estimates of Relative Errors for Average Waiting Time (in s)

STATIONS	CONFIDENCE INTERVAL	ESTIMATE OF RELATIVE ERROR (HALF-WIDTH DIVIDED BY AVERAGE)
Station #10	232 ± 4.32	0.018
Station #9	227 ± 5.02	0.022
Station #8	220 ± 10.49	0.047
Station #7	228 ± 7.06	0.030
Station #6	161 ± 0.81	0.005
Station #5	217 ± 26.86	0.123
Station #4	223 ± 9.01	0.040
Station #3	186 ± 2.77	0.014
Station #2	221 ± 4.45	0.020
Station #1	211 ± 4.82	0.022

Table 8.7 Average Travel Times of Passengers (in s)

	NUMBER OF REPLICATIONS = 10			NUMBER OF REPLICATIONS = 217		
STATION	AVERAGE	HALF-WIDTH	RELATIVE ERROR	AVERAGE	HALF-WIDTH	RELATIVE ERROR
Station #10	585	1.58	0.002	586	0.2	0.0003
Station #9	489	1.47	0.003	488	0.32	0.0007
Station #8	445	3.77	0.008	445	0.52	0.0012
Station #7	432	1.18	0.002	433	0.26	0.0006
Station #6	425	1.84	0.004	425	0.24	0.0006
Station #5	408	6.43	0.015	411	0.94	0.0023
Station #4	471	2.22	0.004	472	0.44	0.0009
Station #3	431	1.62	0.003	432	0.27	0.0006
Station #2	555	1.45	0.002	555	0.27	0.0005
Station #1	484	1.42	0.002	484	0.33	0.0007

increasing and decreasing the number of coaches during the day. The decision has been made to implement only some of the recommendations. The others are discarded due to some political concerns such as public perception of metro services would decline if the number of coaches were decreased at some specific time intervals of a day.

The next step of this study is to study optimization via simulation to determine the optimal headways and the number of coaches to be used for each trip. The optimization model should incorporate the cost of trips into the objectives of minimum waiting times and maximum comfort of the passengers, which is measured by the number of passengers per square meter of the train in each trip.

References

Chang, Y.H., Yeh, C.H., and Shen, C.C., 2000. A multi objective model for passenger train services planning: Application to Taiwan's high-speed rail line, *Transportation Research Part B* 34(2):91–106.

Goverde, R.M.P., 1998. *Synchronization Control of Scheduled Train Services to Minimize Passenger Waiting Times*, TRAIL Research School, Delft, the Netherlands, 35pp.

Law, A.M. and Kelton, D.W., 2000. *Simulation Modeling and Analysis*, McGraw Hill, Boston, MA, 768pp.

Li, J.P., 2000. Train station passenger flow study, *Winter Simulation Conference*, Orlando, FL, Vol. 2, pp. 1173–1176.

Martinez, F.E., 2002. Application of SIMAN ARENA discrete event simulation tool in the operational planning of a rail system, PhD Thesis, University of Puerto Rico, Brazil, 81p.

Sheu, J.-W. and Lin, W.-S., 2011. Designing automatic train regulation with energy saving by using dual heuristic programming, *Proceedings of the IEEE/ASME Joint Railroad Conference*, Pueblo, CO, pp. 355–364.

Yalçınkaya, Ö. and Bayhan, G.M., 2009. Modelling and optimization of average travel time for a metro line by simulation and response surface methodology, *European Journal of Operational Research* 196(1):225–233.

9

Productivity Improvement Studies in a Process Industry

A Case Study

N.S. NARAHARI, K.N. SUBRAMANYA, M.N. VIJAYAKUMAR, AND V. PRASHANT

Contents

Productivity improvement strategies and competitiveness are key issues for enterprise development in any country. Organizations across the globe have an important role in underpinning this in their policy work and in providing services to their customers/clients. Productive enterprise development means that enterprises can create both new and additional jobs, which in turn is one of the most concrete and important contributions to national efforts on poverty reduction as per International Labor Organization (ILO). Many people think that the greater the production, the higher the productivity. This is not necessarily true. Production is concerned with the activity of production of goods and services (output), whereas productivity is concerned with the effective utilization of resources in producing goods and/or services (outcome). Outcome-based conscious improvement strategies assist the organizations in managing their resources effectively. If the organizations in a country are not productive enough to sustain their essential products and services, then the country has to depend on other countries for imports. The greater the dependency, the lesser is the political leverage, particularly in times of war and crisis. If the productivity game were played consistently well by each of the countries in the world, we would have lesser problems from inflation, fewer wages, exploitation, and employment. Countries that have a high growth rate and higher level of labor productivity tend to exhibit a higher standard of living. National economic power further depends upon the level and growth rate of its labor productivity. The factors that affect productivity are the investment of capital per worker, research and development, capacity utilization, government regulation, age of plant and equipment, energy cost, workforce mix, ethics, fear about job losses, and union influence.

In this chapter, the prime focus is on framework of productivity measurement and improvement strategies applied to process industries, in particular. The study has been supported by understanding the current system of operations involved in biscuit manufacturing unit and the issues involved for resolution. Various data collection techniques have been used to record the data for analysis. Quantitative data have been collected to model the system. Linear programming model is used for solving the product mix problem identified in the unit. Other factors affecting the productivity including manpower utilization, material utilization, quality control, and production scheduling were also studied, and improvements for the system were suggested. Finally, the recommendation for engineering and process improvement were suggested to the organization and some of these suggestions are implemented and many others are in the process of implementation.

9.1 Review on Productivity Measurement and Improvement Strategies

Productivity measures are broadly classified into three basic categories, that is, partial productivity, total factor productivity, and total productivity. Partial productivity is the ratio of an output to one class of input. Total factor productivity is the ratio of net output to the sum of associated labor and capacity (factors) input. Total productivity is the ratio of net output to the sum of all the input factors. Productivity can be defined in many ways, depending on who is defining it, whether a politician, an economist, an accountant, a manager, or an industrial engineer.

A productivity strategy does matter a lot in achieving business success. It plays a vital role in achieving the objectives and the organizational goals efficiently. No business can produce better results without measuring and improving its productivity. It can provide ample benefits to a business in a short time. For the most part, it includes business identity development, continuous assessment of workforce, increased business flexibility, measuring competitive edge, and revenue generation. Productivity studies analyze technical processes and engineering relationships including how an output can be produced in a specified period of time. Productivity analysis may be seen as an evaluative activity of the performance of any entity (product or services). If the goal is to increase productivity, enterprises can either produce more with the same level of input or maintain the same level of output using fewer inputs. Another alternative can be increasing the output level with a small increase in the input level. There are certain common set of factors affecting the productivity in every organization. The general categories of the factors concerning productivity include the quantity and quality of labor force, product, quality, process, capacity, and the impacting external influences. Resources are important to consider in the assessment of productivity of an entity, which generally include five M's, namely, men, material, machines, methods, and money. In recent times, there is also a mention of seven M's to include management and measures.

The process of productivity analysis involves conducting detailed comparisons of production reports of different periods and the checking of each line step in the report. In other words, the process occurs from not only distinguishing the items found on the report but also determining the data and documents that are relative to the items and elements of production report. Budgeted and actual time sheets, material requisition forms, manufacturing reports, and material and mixing reports are some of the documents that may have certain values useful for productivity analysis. The first two are the value measures, while the third and fourth are the physical measures. Value measurements raise the problem of currency exchange rates, which fluctuate frequently; hence, the physical measures are better. However, even the physical measures raise serious problems with respect to processes, product mix, and quality.

The three basic approaches of measuring unit labor cost are as follows:

1. *Measurement by product*: This approach considers a typical product and compares labor costs from plant to plant in different countries by using actual cost records. The main drawback of this approach is the variations in cost accounting procedures and cost classification and the variations in the degree of detail maintained in costing; however, if firms do maintain records of cost by products, this approach could be promising. The U.S. Department of Commerce and U.S. Tariff Commission recommend this approach.
2. *Measurement by industry*: This approach consists of aggregating the output of all the different products of individual companies or an industry into a combined figure. Significant studies in this area have been carried out in 1959, as reviewed in the technical papers [1,2]. Although this approach is easier to implement because of the availability of data for basic industries in industrialized countries, there are some difficulties, the most serious ones being quality differences and product diversity.

3. *Measurement by all manufacturing industries*: To overcome the problems of methods proposed by authors in [1,2], measurement is done on a combined manufacturing industry basis. Studies using this approach have been reported by Cooper, Baird and Meissner, and Rostas, in papers [3–5]. The U.S. Bureau of Labor Statistics had applied this method to compare the productivities of various countries.

The two measures used for comparing productivity by the Organizations for European Economic Co-operation (OEEC) are gross national product (GNP) per capita and GNP per person employed. The International Comparison Project (ICP) was another new effort in providing international benchmark for productivity. It was a joint undertaking of the United Nations, the World Bank, and the University of Pennsylvania (with additional financial support at various times from the ford foundation and the foreign aid agencies of several countries, as well as cooperations by the statistical agencies of several countries included in the benchmarking study). The ICP provided comparisons of total and per capita gross domestic product (GDP) at market prices. The U.S. Bureau of Labor Statistics, Office of Productivity and Technology, used the GDP per capita and GDP per employed civilian for productivity comparisons.

Two indexes of national production measures are GNP and net national product (NNP). The GNP index includes capital consumption allowances (i.e., reserves for depreciation of fixed durable capital goods, such as structures and equipment, plus accidental damage to fixed capital). The NNP index excludes the capital consumption allowances. For productivity computation, GNP seems to be preferred due to two major reasons. Computational problem in NNP exists in designing conversion of estimates of depreciation from the original cost values to current time values. Replacement of capital values lost through depreciation can generally be deferred so that depreciation allowance can be assumed to a considerable extent in the short run. At national level, productivity measurement efforts have been primarily from economics of scale point of view. Microeconomists have basically concentrated on using index approach for measuring national productivity. Labor productivity index is a measurement of economic growth of a country. Labor productivity measures the amount of goods and services produced with 1 h of labor as input. More specifically, labor productivity measures the amount of real GDP produced by an hour of labor. Growing labor productivity depends on three main factors: investment and savings in physical capital, new technology adoption, and the use of human capital. The calculations of productivity measurement are as shown in the following:

1. $$\text{Labor Productivity} = \frac{\text{(Real GNP origination in private economy or individual sector)}}{\text{(Man hour of all people employed)}} \tag{9.1}$$

2. As per National Bureau of Economic Research (NBER) indexes:

$$\text{Labor Productivity} = \frac{\text{Output}}{\text{Weighted man hour}} \tag{9.2}$$

3. As per Brookings Institute Indexes (BIIs):

$$\text{Total Labor and Capital Productivity} = \frac{\text{Output}}{\text{(Labor} + \text{Capital)}} \tag{9.3}$$

where

Labor input is in man-hours, adjusted for quality change in labor

Capital input is the net stock of structure + plant equipment + inventories + working capital + land

Productivity improvement strategies in an organizational context can use one of the two major approaches including continuous improvement method and breakthrough improvement method.

Continuous improvement methods have been popularized by Japanese manufacturing organizations and have been known as *KAIZEN*. Resource usages in the organization are monitored continuously, opportunities for resource utilization are identified, and productivity improvement ideas are implemented in manufacturing plant. These strategies require the involvement of shop-floor persons in identifying the element of waste (*MUDA*) and the project approach to eliminate waste in a systematic manner. The changes are introduced in the method of production and the use of resources leading to overall improvement in the plant productivity. These changes are then institutionalized to maintain the gains in the system. These types of approaches are commonly adopted in developing and developed economies to improve manufacturing practices continuously with a focus on product, process, and system-level changes. Continuous improvement methodologies bring in incremental changes in organizational productivity in factory floors, offices, retail outlets, service units, institutions, or any other entity being investigated. Continuous productivity improvement strategies is a long-term approach to bring in a culture of a *war on waste* by increasing the waste hidden in the factories or the service units that engage in the conversion process of converting inputs into useful outputs in terms of products or services. Continuous improvement strategies are necessary for the standardized and well-documented organizational processes and business processes. A system approach to problem solving lies at the heart of continuous improvement methods. This approach demands a higher-level utilization of manufacturing and deploys industrial engineering tools and techniques to identify and eliminate the gap between the targeted and achievable productivity levels. The use of lean philosophy and lean methodologies has been popularized by various practitioners and preachers of continuous improvement over the last three to four decades. This approach has root in American manufacturing plants and has spread to other parts of globe as being key to building globally competitive productivity improvement strategies.

The second set of approaches relies on innovation or breakthrough in the technology and methodologies adapted. The innovation approach brings radical ideas and generates a possible quantum leap in productivity measures. The breakthrough approach for productivity enhancement will lead to overhauling manufacturing practices and adopt radical questioning of the current status quo in the system. In the face of stiff competition, the breakthrough innovations in the organization can provide edge to individual organizations as a whole or for the entire community. The breakthrough approaches require a culture of research and development in science and technology and investment in capital to generate innovative ideas that can be game changer in the market. The use of game changers as strategy for productivity measurements and improvement can benefit organizations to face the competition from external environment and gain popularity in an Internet-worked global enterprise model of doing business. The order or scale of improvement through breakthrough innovation can be much more on a large scale compared to the philosophy of increasing productivity improvement method [6–11].

The earlier works in the area of productivity provide better insight into the working mechanism in industries and improvements achieved in the organizations. In the production environment, it is well known that batch process manufacturing is different from discrete manufacturing. Finished goods are created through a mix of continuous and batch manufacturing processes that makes multiple intermediate products that are then converted into hundreds of finished goods. Production recipes with both convergent and divergent material flows are very different from bills of material. Cleaning procedures are mandatory, more frequent, and often more disruptive than equipment maintenance in discrete manufacturing. Filling and packaging lines are generally more automated, and batch control and traceability are enforced throughout the entire production process. One result of all of these differences is that while there are many more stock keeping units (SKUs) in discrete manufacturing, production planning and detailed scheduling are generally more difficult in batch process manufacturing plants than in work-center-based manufacturing or line-based assembly processes. This white paper reviews the planning and scheduling challenges faced by many batch process manufacturing companies and provides information about new approaches that are making it easier to manage batch operations to achieve new levels of performance, control, and flexibility [12]. Globalization is posing several challenges to the manufacturing sector. Design and operation of manufacturing systems are of great economic importance. Factory performance remains unpredictable, in spite of the considerable literature on manufacturing productivity improvement and the long history of manufacturing, as there is no widespread agreement on how it can best be performed. Productivity measurement and improvement go hand in hand, because one cannot improve what one cannot measure. The review of literature on manufacturing systems productivity measurement and improvement has been summarized under four categories: they are operations research–based methods, system analysis–based methods, continuous improvement methods, and performance metrics–based methods.

A survey of commercial tools available to measure manufacturing system performance is also performed. The review indicates that quantitative metrics for measuring factory-level productivity and for performing factory-level diagnostics (bottleneck detection, hidden capacity identification) are lacking. To address this gap, a factory-level effectiveness metrics–based productivity measurement and diagnostic methodology is proposed [13]. A system approach to the problem of tackling low numerical control (N.C.) machine utilization has been developed following a study in a company of the productive efficiency of a group of N.C. machines, which were found to have poor levels of utilization. The many reasons for the low utilization were analyzed. To improve the utilization of the machines in a way that was likely to lead to a lasting improvement, methods of improving utilization were developed, which looked at the machines within the engineering and management systems in which they operated [14].

9.2 Productivity Measurement and Improvement in Process Industries

In a process industry, the mode of production is batch manufacturing. Batch manufacturing involves the passage of the material through the various stages of production with chemical and physical transformation steps involved in the conversion of raw material into the final finished product. Examples for batch manufacturing include manufacturing in core industries such as steel, coal, mining, and pharmacy sector including production of medicines, food, and confectioneries such as chocolates, biscuits, and a variety of food and spice in a packaged form. Productivity measurement and improvement strategies in batch manufacturing and the process industries distinguish themselves from the strategies used in mass production industries such as automotive component and assembly manufacture. Productivity losses in process industries arise due to yield losses and production stoppages due to maintenance requirements and the lack of feed material for the continuous production runs.

The productivity tracking system in a process manufacturing unit, such as the biscuit manufacturing unit considered for study in this chapter, involves the monitoring of the production rates, resources consumed during the batch manufacturing, cracking, wastages of material, and monitoring yield losses. In addition, productivity can be impacted as a result of the system of maintenance practices. Preventive maintenance of equipments and machinery used in batch production can lead to minimization of downtime of the machinery. Prescription of optimal preventive maintenance schedules for the machinery can lead to overall efficiency of the batch production system. Defect minimization and reduction of yield losses is an important part of the overall productivity measurement and improvement in a batch manufacturing setup.

9.3 Case Study on Productivity Improvement in a Process Industry

The case study presented in this chapter demonstrates an integrated production/operations planning, quality monitoring, and control, with time as a parameter to meet the delivery. The challenges of productivity measurement and improvement have been tackled through a scientific investigation of the causes leading to poor production and utilization of resources. This work is focused on the productivity improvement through incremental changes in the areas of operations, quality, and maintenance practices in a biscuit manufacturing unit. The investigation into current practices has revealed opportunities for improvement in overall organizational productivity. In the current system, the technology being adopted involved semiautomation with involvement of machinery and a major component of labor. The machinery was quite old and a program of modernization would mean a major investment and capital expenditure decision, which the unit is not in a position to afford financially. Hence, the focus was laid on methodology rather than the technology of productivity improvement. As the process that has been studied in this case study is labor intensive and is partially automated, the process related to biscuit manufacturing has been analyzed for productivity improvement through continuous improvement as a preferred strategy.

9.3.1 Current System of Operations Involved in Biscuit Manufacturing Unit

The study was carried out at M/s. XYZ Foods Pvt. Ltd., which is a unit committed to the manufacture and supply of 700 metric tons (MT) of biscuits (in all the four varieties) per month. Presently, the unit is able to manufacture only 600 MT of biscuits (in all the four

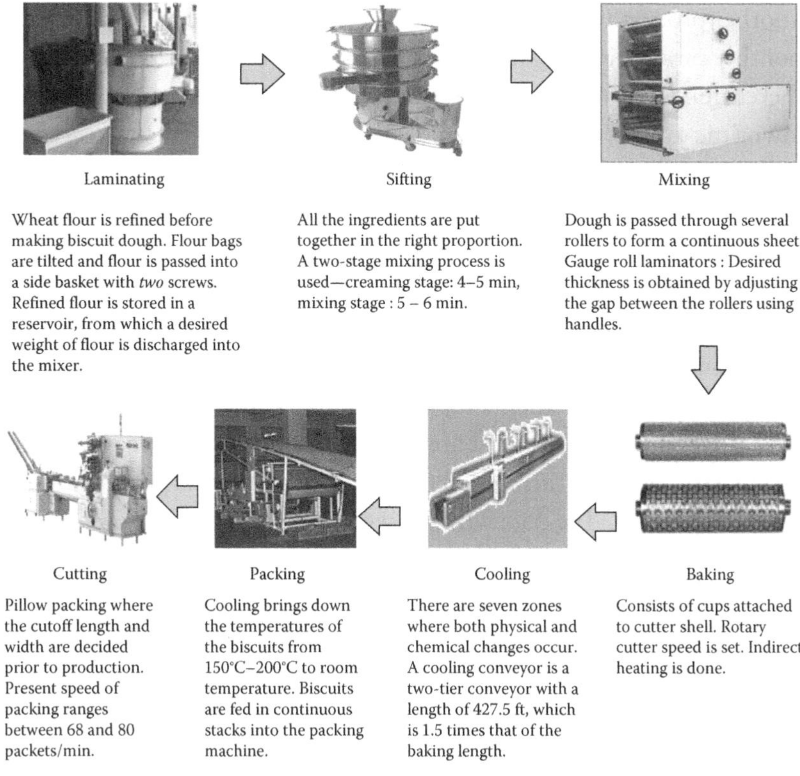

Figure 9.1 Chart depicting operations and processes involved in biscuit manufacturing.

varieties) per month. The process of manufacturing biscuits can be classified into six main stages: sifting, mixing, forming, baking, cooling, and packing.

A brief description of the processes involved is provided (Figure 9.1).

9.3.2 Problem Identification and Data Collection

There is a noticeable variance between the actual output and the desired output. Due to this, the unit had a threat of losing the orders to its competitors in the biscuit manufacturing business. The operation plan was based on little analysis or intuitive judgment and past data. After studying the present system of operations, it was observed that there was slackness in the overall production. This was due to frequent and unscheduled changeovers from one variety of biscuits to another.

Another process was the high defect rate, this being 3.5% of the total monthly production in all the four varieties of biscuits manufactured. The current maintenance operations resulted in an average monthly downtime of 12%. Further, manual operations in sifting, mixing, and laminating section caused a lot of fatigue to the workers leading to reduced overall productivity levels. To address these problems, data collection was done through primary and secondary sources. The company receives the data on the quantity to be manufactured under four varieties every Friday. Therefore, the department has to plan its activities and schedule the production of the four varieties of biscuits. Accordingly, the ingredients for all the four varieties are supplied well in advance. The percentage share of each biscuit variety was collected through the manufacturing logbooks and by interacting with the personnel concerned. The data with respect to the quality of biscuits manufactured, overall plant downtime, the number of men and women involved in the total production process, etc. were recorded from the regular reports of the company for a period of 6 months. The quantity produced per shift in terms of MT, the number of batches, and the various inputs used in the process were collected from the production and planning report generated on a daily basis. Daily Quality Index (DQI) report prepared the details of the inspection carried out on biscuits on random basis in all the three shifts. The inspection is carried out for quality characteristics related to taste, appearance, smell, etc. This report provides the inspection of wrapping sheet rolls and BOPP films used for packing the biscuits in terms of appearance and any other defects. This report helps in computing the DQI using the average of packing index (PI) and biscuit index (BI) using a grading system of excellent, very good, good, and unsatisfactory based on the requirements of the specifications. The downtime data were recorded between every shift. The following major causes for plant downtime were identified:

- *Mechanical*: Mechanical problems include frequent breakdown of gauge roller chain drive mechanism, sudden failure of oven burners due to choking, and improper functioning of the flour sieve.
- *Electrical*: Electrical problems include frequent and erratic power failures, which are further aggravated due to the

inadequate in-house generator capacity and also because of frequent short circuits in the electricity board.
- *Operational*: Operational problems include faulty mixing due to worker's negligence and inadequate availability of trays at the packing section.

Daily report is a summary report that gives *on-date* and *up-to-date* information about the production and dispatch in terms of the number of cartons produced and dispatched and MT of biscuits produced and dispatched. Also, the fuel (in L) and power (in kW) utilized in the process of manufacturing, packaging, and dispatching are figured in this report. The number of men and women deployed was also collected from the daily report.

A flow process chart was developed to record the job content of each worker in the different regions of the shop floor to understand the level of utilization of men in the plant. The entire work was divided into operation, transportation, inspection, delay, and idle elements. It helped in tracking unnecessary movements and identifying delay times in each element of the job.

The quantity of each ingredient available was noted down and highlighted in case of shortages. Also, the wastage of different materials like dough, wheat flour, and leavening agents in kg was recorded. The amount of rework and rejects was calculated by segregating the nature of defective biscuits and thereby finding out the percentage of each. The working capacities of each machine were noted down. The total breakup of running time and idle time was estimated, and reasons for the nonproductive time elements were highlighted with their causes. The detail of all the changeover times was gained from the downtime reports and is as shown in Tables 9.1 and 9.2.

9.3.3 Linear Programming Model for Product Mix Problem

After collecting the data from various reports in the company and preliminary analysis, it was decided to use a mathematical approach for achieving the optimal product variety combinations. After reviewing the literature, it was found that linear programming approach for determining optimal production plan was suitable in the process industry context. To determine the optimal product manufacturing mix for the

Table 9.1 Flow Process Chart Showing Operations

DETAILS OF THE METHOD	OPERATION	TRANSPORT	INSPECTION	DELAY	IDLE	TIME (s)	NOTES
Stands in stores					*		Involves lot of worker fatigue as work is done manually
Transfers one sack onto the trolley	*					10	
Moves with trolley toward the sifter		*				35	
Transfers sack onto the sifter arm step	*					10	
Moves to the stores with empty trolley		*				30	
Repeats (2–5), 20–30 times in 1 h						85	

Note: * indicates activities carried out at different stages of production.

Table 9.2 Flow Process Chart Showing Operations: Sifter Region (Worker 2)

DETAILS OF THE METHOD	OPERATION	TRANSPORT	INSPECTION	DELAY	IDLE	TIME (s)	NOTES
Stands next of sifter arm 1					*		
Removes thread and opens the sack	*					10	
Unloads the sack partially into sifter arm 1	*					15	
Pauses for the arm to get half empty				*		10	
Unloads the sack completely	*					10	
Moves the sifter arm 2		*				5	
Waits for the next sack to arrive				*		5	
Removes the thread and opens the sack	*					10	
Unloads sack partially into the sifter arm 2	*					15	
Pauses for the arm to get half empty				*		10	
Unloads the sack completely	*					10	
Moves the sifter arm 1		*				5	
Repeats (2–12), 10–15 times in 1 h						105	

Note: * indicates activities carried out at different stages of production.

Table 9.3 Standard vs. Achieved Production Rate

VARIETY	ACHIEVED QTY. PER MONTH (MT)	ACHIEVED QTY. PER SHIFT (MT)	STANDARD PER SHIFT (MT)
Variety I	240	8.0	9.0
Variety II	200	6.6	9.0
Variety III	120	4.0	6.5
Variety IV	40	1.3	6.0
Total	600	19.9	30.6

four varieties, a linear programming model was formulated. The model was built keeping in mind the different constraints that include the capacity constraint, manpower constraint, and material constraint during each production month. The standards set by M/s. ABC Industries Ltd., which gave suborders, was followed by M/s. XYZ Foods Pvt. Ltd. The monthly production distribution details of all the four varieties of biscuits are as shown in Table 9.3. It consists of the variety of biscuits quantity per month MT and standard output per shift. It is noticed that variety I is manufactured the most and variety IV is manufactured the least. This production schedule was suboptimal.

9.3.3.1 Identifying and Defining Decision Variables

Let x_1 be the quantity of biscuits of variety I to be produced per month.

Let x_2 be the quantity of biscuits of variety II to be produced per month.

Let x_3 be the quantity of biscuits of variety III to be produced per month.

Let x_4 be the quantity of biscuits of variety IV to be produced per month.

9.3.3.2 Objective Function The objective function was to maximize the overall production of the manufacturing line across all the varieties of biscuits. This is expressed mathematically as

$$\text{Max}\,Z = 240x_1 + 200x_2 + 120x_3 + 40x_4 \tag{9.4}$$

9.3.3.3 Constraints Physical constraints and conceptual constraints relate to the availability of raw material supply, manpower, and capacity limitation on the critical process machinery.

Material constraints:

$$\text{Wheat flour: } 564.66x_1 + 564.66x_2 + 564.66x_3 + 564.66x_4 \le 5{,}31{,}250 \quad (9.5)$$

$$\text{Sugar solution: } 209.57x_1 + 214.45x_2 + 266.66x_3 + 242.22x_4 \le 7{,}50{,}000 \quad (9.6)$$

$$\text{FCCM: } 4.44x_1 + 8.89x_2 + 8.89x_3 + 0x_4 \le 10{,}000 \quad (9.7)$$

$$\text{Palm oil: } 72.18x_1 + 62.53x_2 + 73.24x_3 + 56x_4 \le 46{,}000 \quad (9.8)$$

$$\text{Maize starch: } 0x_1 + 22.22x_2 + 0x_3 + 22.22x_4 \le 10{,}000 \quad (9.9)$$

$$\text{Arrow root starch: } 0x_1 + 0.93x_2 + 0x_3 + 0x_4 \le 1{,}000 \quad (9.10)$$

$$\text{SMP solution: } 0x_1 + 13.33x_2 + 33.18x_3 + 13.33x_4 \le 32{,}000 \quad (9.11)$$

$$\text{GMS paste: } 0x_1 + 10x_2 + 6.2x_3 + 0x_4 \le 4{,}000 \quad (9.12)$$

$$\text{Ammonia: } 7.78x_1 + 7.78x_2 + 7.78x_3 + 7.78x_4 \le 10{,}000 \quad (9.13)$$

$$\text{Soda: } 2.67x_1 + 2.67x_2 + 2.77x_3 + 3.33x_4 \le 10{,}000 \quad (9.14)$$

$$\text{Salt: } 4x_1 + 3.11x_2 + 3.11x_3 + 4.44x_4 \le 10{,}000 \quad (9.15)$$

$$\text{SMBS: } 3.33x_1 + 2x_2 + 3.04x_3 + 1.33x_4 \le 3{,}000 \quad (9.16)$$

$$\text{Biscuit dust: } 13.33x_1 + 13.33x_2 + 22.22x_3 + 22.22x_4 \le 10{,}000 \quad (9.17)$$

Manpower constraints: The manpower constraints were expressed in terms of man-hour available in a production:

$$35.55x_1 + 35.56x_2 + 49.23x_3 + 53.33x_4 \le 23{,}685 \quad (9.18)$$

Capacity constraints: The capacity constraints were also considered in view of the limit of the production units that can be processed; the production rates were different at different stages of manufacturing:

$$\text{Laminator: } 1x_1 + 1x_2 + 1x_3 + 1x_4 \le 720 \quad (9.19)$$

$$\text{Packing M/C: } 1x_1 + 1x_2 + 1x_3 + 1x_4 \le 800 \quad (9.20)$$

$$\text{Cutter: } 1x_1 + 1x_2 + 1x_3 + 1x_4 \le 1182 \quad (9.21)$$

$$\text{Mixer: } 1x_1 + 1x_2 + 1x_3 + 1x_4 \leq 845 \quad (9.22)$$

$$\text{Oven: } x_1 + 1x_2 + 1x_3 + 1x_4 \leq 1182 \quad (9.23)$$

Nonnegativity constraint: $x_1 \geq 0$, $x_2 \geq 0$, $x_3 \geq 0$, $x_4 \geq 0$

9.3.3.4 Output The output obtained from STORM software provides the gross details of the linear programming problem (LPP) optimization model. In this problem, the number of variables chosen is 4, and there are 24 constraints in the LPP model. The starting solution is at the origin, and the objective function is to maximize the total monthly production output from the four varieties of biscuits. The output obtained lists the optimal product mix by providing the values of the objective function coefficients and the resource consumed for the production of the four varieties of biscuits. The constraints considered in the problem include the material constraints and labor and machine constraints. This output also provides information on the lower bound and the upper bound. However, while solving, the bounds were not stipulated in the formulation. Objective function value obtained is Rs. 113,614.90. Tables 9.4 and 9.5 provide the optimization of product mix and range and target values of variables. Also, Table 9.6 provides the sensitivity analysis

Table 9.4 Optimization of Product Mix

VARIABLES	MONTHLY OUTPUT IN MT
x_1	240.0056259
x_2	200
x_3	120
x_4	40
Objective total (Z max)	600.0056259
Objective function value	113614.9

Table 9.5 Range and the Target Values of Variables

VARIABLES	VALUE	LOWER LIMIT	UPPER LIMIT	TARGET VALUE
x_1	240.0056	240	240.0619	240.0056
x_2	200	200	200.0619	200
x_3	120	120	120.0447	120
x_4	40	40	40.0413	40

Table 9.6 Sensitivity Analysis of Cost Coefficients

S.L. NO.	VARIABLE	CURRENT COEFFICIENT	ALLOWABLE MINIMUM	ALLOWABLE MAXIMUM
1.	VAR I	240.0000	200.0000	∞
2.	VAR II	200.0000	$-\infty$	240.0000
3.	VAR III	120.0000	$-\infty$	332.3544
4.	VAR IV	40.0000	$-\infty$	360.0338

of the current value of the b_j coefficients and also gives the allowable minimum and maximum values. As long as the b_j values remain within the allowable minimum and maximum, the current optimal solution is valid.

The recommendation given to the unit was in terms of the optimal production of the four varieties of biscuit and the adoption of batching strategies to mix the production of the four varieties in the monthly production schedule. The LPP as a modeling structure has relevance to the volume–variety batch production situation existing in the biscuit manufacturing unit.

9.3.4 Other Factors Affecting Productivity in the Biscuit Manufacturing Division

In addition to the optimization of a variety of biscuits produced, the study also focused on other factors that are studied for enhancing productivity. The improvements of the following factors are done using industrial engineering tools and techniques. The current and proposed systems are compared for showing the improvement in productivity:

- *Manpower utilization*: Presently, there are 35 workers in the manufacturing division. The present system was observed using the flow process chart. Interview and online observations indicated that the number of workers in each stage of the manufacturing process required is less than the present manpower. Table 9.7 depicts the manpower required during the various stages of production in the present and proposed system. Figure 9.2 depicts the value-added and non-value-added activities in the various processes. There is a reduction of 30% manpower in the proposed system.

Table 9.7 Manpower Utilization in the Present and Proposed Method

NO. OF EMPLOYEES	SIFTER	PREMIXER	MIXER	BISCUIT GRINDING	LAMINATOR	GAUGE ROLL	CUTTER	COOLING CONVEY	PACKING
Present	2	3	2	2	1	2	2	1	20
Proposed	1	2	1	1	1	1	1	1	15
Savings	1	1	1	1	0	1	1	0	5

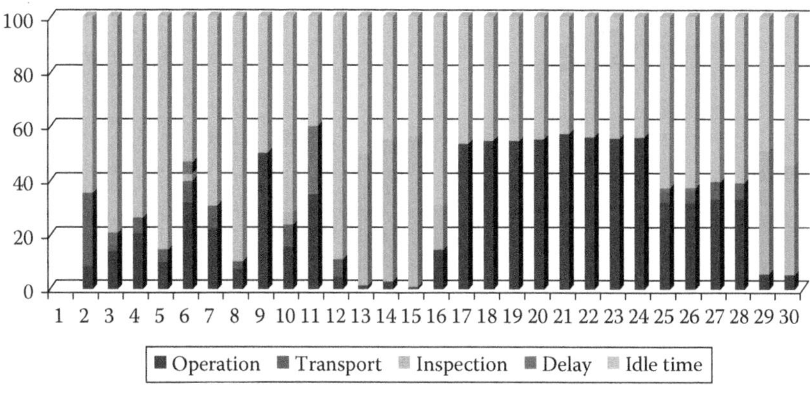

Figure 9.2 Distribution of productive and nonproductive element of work.

Table 9.8 Material Utilization in the Present and Proposed Methods

PRODUCT VARIETIES	PRESENT		PROPOSED	
	GROSS YIELD (%)	NET YIELD (%)	GROSS YIELD (%)	NET YIELD (%)
Varieties I and II	84	81	90	87.3
Varieties III and IV	83	80	90	87.3

- *Material utilization*: The current defect rate is 3.5%. This is reduced to 3% by making possible changes in the different stages of the manufacturing process of biscuits. Hence, the net yield has improved by 6.8% and the results are depicted in Table 9.8. Also, Figure 9.3 indicates the percentage of wastes involved in various processes of biscuit manufacturing.

  While studying the processes, the assignable causes were also identified in various processes, which caused defects in the processes. Table 9.9 provides the issues identified, assignable causes, and possible solutions for improvement.
- *Machine utilization and quality control*: The bottleneck analysis emphasized the fact that laminator was the bottleneck. The proposal of a new laminator is suggested considering cost–benefit analysis. This improves output production from existing 1 to 1.2 ton/h, which is substantial, since the processes are continuous. A sample graph as shown in Figure 9.4 indicates control limits to be monitored for better

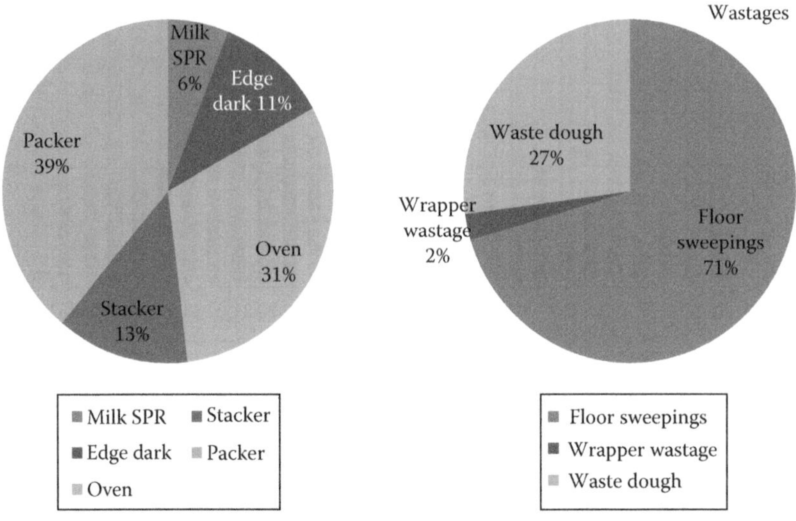

Figure 9.3 Percentage of wastages in various processes.

Table 9.9 Issues, Causes, and Solutions for Improvements

ISSUES	ASSIGNABLE CAUSES	SOLUTION
Packer defects	• Manual changeover of the wrapper rolls • Frequent mechanical problems in the packing machine	• Upgradation of wrapper roller change equipment • Arrangement of biscuits into packing machine using U-shaped chute design
Oven defects	• Frequent and erratic power failures causes burning of biscuits	• Need for quicker takeover by the generator during power failure
Stacker	• Improper arrangement and handling of the biscuits	• Stacker redesign
Higher wrapper, floor sweepings, and waste dough	• Manual movement of raw material and dough in sifter and mixer region • No systematic attack on the problem and wastage that is accepted as routine	• Tray conveyor for movement of raw material in sifter region and lift mechanism in the mixer region • Ergonomic design of the trolley used in the mixer region

production for variety I type of biscuit. In order to monitor the reason for out of control values for other varieties, charts were drawn.

The present defect rate was 3.5%, and after the necessary changes suggested during the study, it was estimated that

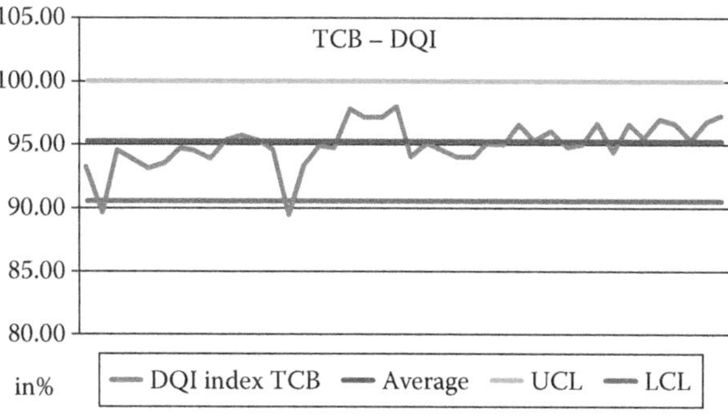

Figure 9.4 Quality indicator for variety I. *Note:* TCB, Tiger Chai Biskoot; DQI, Daily Quality Index.

the defect rate would fall to a reasonable 3% leading to an improvement of 0.5%. This has enhanced productivity.

- *Production scheduling*: One of the major lacunae in the system was higher time consumption during the changeover. With the revised scheduling of the products considered during production planning, the time consumed was saved to the extent of 40%. The downtime per month was about 200 min. Using the changeover schedules, improving the mechanical and operational problems, the downtime has been reduced to 150 min/month.

9.3.5 Summary of Results and Recommendations

The optimal solution for the manufacturing of different varieties of biscuits matched with those given by the company. While formulating the model, optimizing the resources in order to improve the productivity of the plant was the concern, which was addressed through improving utilization of men for efficient running of the plant, improving material utilization by improving the net yield by minimizing the number of internal rejections at various stages of production process, and enhancing the machine utilization by reducing the downtime caused due to frequent and sudden breakdown of a few of the critical machines in the exit. To reduce the downtime and improve the productive time of the plant, a proper inspection plan was carried out, after which the breakdown maintenance was replaced by preventive maintenance.

Table 9.10 Measures of Performance for Overall Productivity Improvement

MEASURES	PRESENT	PROPOSED	IMPROVEMENT (%)
Defect rate	3.5%	3%	0.5
Average changeover time	35 min	20 min	40
Machine utilization	1 ton/h	1.2 ton/h	20
Manpower utilization	29 persons	20 persons	31
Material Utilization	80.5%	87.3%	6.8
Downtime	200 min/month	150 min/month	25

As a result of the several ideas suggested by the team, following improvement in the key performance was recommended after a trial implementation. These measures of performance are helpful for the company in specific and process industry in general to enhance productivity of the organization to be more competitive in the market in terms of quality, price, and delivery. Table 9.10 depicts the snapshot of the case study and serves as a ready reckoned for decision makers for continuous improvement.

The study and trial implementation also takes into consideration the subjective and engineering factors for improvement in the productivity. Table 9.11 indicates the improvements in terms of physical design and processes.

A trial implementation of some of the suggestions recommended has already been undertaken. The other suggestions are under consideration. The management of the company has set up a cross functional team for long-term suggestions made by the group.

9.4 Conclusion

This study has considered productivity as a center theme and has dealt with the productivity concept in international, national, organizational, and shop-floor/factory-floor level. The issues that were in batch manufacturing production system, considering the case of biscuit manufacturing, have been discussed. The productivity improvement and resource optimization ideas have been exemplified. Empirical data have been analyzed to identify problem area in the existing production system of biscuit manufacturing. Mathematical modeling approach for maximizing total production within optimal product mix of variety in a week production targets has been used.

Table 9.11 Recommendations for Engineering and Process Improvements

PROCESSES	SOLUTIONS FOR IMPROVEMENTS
Sifter	Instead of a person doing the work manually and creating lots of maida wastage, tray conveyor with automatic tilts is recommended, which dumps maida into hopper connecting two tubes for each arm of sifter. Through the conveyor system, the arms transfer the maida directly into the sifter with less wastage.
Mixer	Premixing of ingredients is done manually at present. The use of a premixer will result in shorter preparation time of dough in the mixer. Raw material addition and transfer of liquids for mixing can be automated with help of programmable logic controller (PLC)–based system along with weighing and batching done with automated system. Weighing is done through load cells and electropneumatic valves that can be integrated with PLCs. Mixers with the help of timers and software can adjust mixing timing as per dough consistency. The trolley that moves the material from mixer to laminator is not proper in terms of taking the entire load and when wastage occurs. Hence, redesign of trolley with greater width and greater height from the ground with lift mechanism will help in carrying more load. Regular maintenance and cleaning process after each shift must be made as a routine practice.
Laminator	Presently, the angle swing is high. Reduction of the oscillator spread by reducing the angle of swing reduces the wastage of dough in the sides of the conveyor and material accumulation also reduced at a point. Increasing the capacity of laminator or buying a new laminator with increased capacity will increase the output.
Cutter	Using sharp cutter blades and inspecting the quality of dough (hard/soft mix) lead to avoidance of sticking of dough on the cutter.
Oven	The use of uninterruptable power supply (UPS) will take over the oven quicker than the present generator avoiding the burnt biscuit formation. Inspection of burners at regular periodic intervals is necessary.
Cooling conveyor	Placing tray boxes at the points where the conveyor belt takes turns and places where biscuit accumulation is high, angle plates are to be placed at turns when biscuit comes out of the oven. This avoids wastage of biscuits. The length of conveyor belt 1 is to be increased, which goes over belt 2 so that the material does not fall down.

The material wastages have been identified and causes attacked to improve the yield from the process. Maintenance systems have been analyzed to measure downtime losses. In conclusion, an integrated framework of improving productivity, focusing on planning and implementing production system, treating quality issues, and solving them in addition to these maintenance issues has been analyzed.

This chapter focuses on productivity studies in process industries, in general, and biscuit manufacturing firm, in particular. The study involves continuous production system as applied to biscuit manufacturing firm, which includes processes on laminating, sifting, mixing, baking, cooling, packing, and cutting for manufacturing quality biscuits. There is a tremendous scope for resource optimization in the continuous production systems as well which can be taken up as scope for further improvements in the plant. In addition, constraints that were formulated were basically considering the internal resource requirements and very limited focus on external constraints such as market requirements. Optimization models can be developed by introducing additional external constraints. Thus, there would be a need to continuously monitor and find better ways of planning and executing the biscuit manufacturing processes through technological improvements and innovations for continuous productivity and quality improvements in the area of biscuit manufacturing.

References

1. Paige, D. and G. Bombach. *A Comparison of National Output and Productivity of the United Kingdom and United States*. Organization for European Economic Cooperation, Paris, France, 1959.
2. Williams, F.M. and E.I. Eaton. Payment for labor and foreign trade. 49, 584–601, 1959.
3. Cooper, R.N. The competitive position of the United States. In S.E. Harris, ed., *The Dollar in Crisis*. Harcourt, Brace & World, New York, 1961, pp. 1579–1593.
4. Baird, M.K. and F. Meissner. *Wage Cost Abroad*. Stanford Research Institute, Menlo Park, CA, 1961.
5. Rostas, L. *Comparative Productivity in British and American Industry, Occasional Papers XIII*. National Institute of Economics and Social Research, Cambridge University Press, London, U.K., 1978.
6. Riggs, J.L., L.L. Bethel, F.S. Atwater, G.H. Smith, and H.A. Stackman Jr. *Industrial Organization and Management*. McGraw-Hill, New York, 1979.
7. Siegel, I.H. *Company Productivity, Measurement for Improvement*. The W.E. Upjohn Institute for Employment Research, Kalamazoo, MI, 1980.
8. Sumanth, D.J. Productivity measurement and evaluation models for manufacturing companies, PhD dissertation. Illusion Institute of Technology, Chicago, IL. University Microfilms International, Ann Arbor, MI, 1979, pp. 164–174.
9. Adam, N.R. and A. Dogramaci, eds. *Productivity Analysis at Organizational Level*. Kluwer, Boston, MA, 1980.
10. Davis, H.S. *Productivity Accounting*. University of Pennsylvania Press, Philadelphia, PA, 1995.
11. Atsushi, K. Product market competition and productivity in the Indian manufacturing industry. *Studies* 45(10), November 2009.

12. Focacci, F. *Improving Plant Performance and Flexibility in Batch Process Manufacturing: With an Example from the Food and Beverage Industry.* IBM ILOG Supply Chain Applications, Application and Integration Middleware Software, IBM Software Group, Vol. 1, No. 4, 2009, pp. 283–301.
13. Muthiah, K.M.N. and S.H. Huang. *A Review of Literature on Manufacturing Systems Productivity Measurement and Improvement.* University of Cincinnati, Cincinnati, OH.
14. Kilmartina, B.R. and R.G. Hannamba, *An In-Company Study of N.C. Machine Utilization and Its Improvement.* UMIST, Manchester, U.K.

10

Modeling Inventory Dynamics

The Case of Frenudco

IJAZ YUSUF, TASHFEEN M. AZHAR, AND SARWAR M. AZHAR

Contents

10.1 Introduction

Although the discipline of industrial dynamics, commonly known as system dynamics, was developed some 60 years ago, it is still viewed in the East as a new technology for policy design in management and planning (Richardson and Pugh, 1981). Broadly speaking, the systems thinking, feedback notion, and control engineering are applied to economics, business, and organizational systems (Towill, 1996) to gain a better insight into the problems under study. The objective of the chapter is to simulate the dynamics of the inventory management, narrate the basic concepts of system dynamics, and design egalitarian policies.

Systems thinking, business dynamics, industrial dynamics, systems approach, system dynamics, and systems theory are a few terms commonly attached to this field that masterly attempts to combine the key concepts like feedback controls, mutual causality, nonlinearity in the functions, cybernetics, complexity, counterintuitive behavior, deviation correcting and deviation amplifying processes like goal seeking, external resource production process, and many more with the organizational systems (Sterman, 2007). The creator of system dynamics, Forrester (1980), defines it as a methodology that starts with important problems, comes to understand the structures that produce undesirable symptoms, and moves on to find changes in structure and policy that would make a system better behaved. System dynamics, being a blend of knowledge of control engineering, cybernetics, and organizational theory, is a guiding philosophy to analyze the dynamic behavior of a model in terms of its feedback mechanisms. It is a method of dealing with questions about dynamic tendencies of complex systems, that is, the behavior patterns they generate over time (Meadows et al., 1994). Systems thinking, in practice, is a continuum of activities which ranges from conceptual to technical (Richmond, 1993). During the first phase, reference mode and influence diagram are constructed as a presimulation model (Howich et al., 2006). The modeling protocol is described by Rabia and Yusuf (2007) is as follows:

PHASES	DESCRIPTION	NATURE
Phase 1	Problem definition	Conceptual
	System perspective	Conceptual
	Reference mode	Conceptual
Phase 2	System conceptualization	Conceptual
	Causal loop diagram	Conceptual
	Influence diagram	Conceptual
	Dynamic hypothesis	Conceptual
Phase 3	Block diagram	Technical
	Stock-flow diagram	Technical
	Model representation	Technical
	Model structure	Technical
Phase 4	Equation writing	Technical
	Model simulation	Technical
	Model behavior and evaluation	Technical
	Scenario testing and policy design	Technical

System dynamics model is an art one can learn through apprenticeship and dedicated practice. Playing with the model to find the equilibrium state on different sectorial basis, repeated experimentation with the system, testing assumptions, evaluating current scenarios, and altering management policies are the real benefits of the model. The purpose is to gain understanding so that the problem to which the model is addressed could be solved or minimized. How the model will look like in the end depends a lot on the selection of dynamic variables, transparent goal, clear and unambiguous description of the situation, and closed boundary of the model. Without a purpose, there can be no answer to the question of what system components are important (Forrester, 1968). Without the system boundary, it is impossible to define the modes of behavior that are created by the interaction of system components (Weil, 2007).

10.1.1 Positive and Negative Causal Loops

Decision stream and ability to analyze the behavior of models rely heavily on simple models linking feedback loops to system behavior (Oliva et al., 2006). Influence diagram and simple causal loops lead to conceptualization of real-life scenario during the model formulation. Causal loop is a closed sequence of causes and effects, a closed path of action and information. A decision is based on the observed state (level) of the system. The decision produces action that alters the state (level) of system, and a new state (level) gives rise to new information as the input to further decisions. Behavior of the system is the result of the interaction of positive and negative feedback control loops (Lane, 2007).

A positive loop generates runaway growth or collapse behavior that creates virtuous or vicious cycles: a change in one variable brings change in the next variable. Compounding and reinforcement are the outcomes of the positive loop. Negative feedback loop is a goal-seeking loop that maintains the status quo and generates honing-in behavior. Negative loops resist the change, push them one way, and they come back while maintaining the equilibrium state. System thinkers define it as follows: "When a feedback loop response to a variable opposes the original perturbation, the loop is negative or goal seeking. The definition of a negative loop is usually interpreted to mean that…a change in one element is propagated around the circle until it comes back to change that element in a direction opposite to the initial change" (Meadows, 1994).

10.1.2 Level and Rate Variables

Phase 3 of the modeling protocol starts with the block and flow diagram that comprises level and rate variables. Associated variables like auxiliary or convertor, table function, exogenous variable, source or sink of material, material and information flows, and constants to represent parametric changes are building blocks to complete the block diagram. Feedback structure can be portrayed by equations or stock and flow diagrams (Richardson, 1986).

Levels reflect on conditions within the system at a point in time. Stocks are what we would see if we suddenly freeze the activity within the system. Levels are just like the bathtubs in the sense that they accumulate or collect flows. Levels are said to be stocks in economics and integrations in mathematics, and in engineering, they are referred to as system state variables. Rates represent the stream of activity associated with particular stocks. Flows are depicted by a pipe that represents the conduit through which goal-seeking activities flow. They are just as valves that feed and drain stock (Yusuf and Ramish, 2011). Rates depend only on the values of the level variables. Policy statements in the system define how the existing conditions of the system produce a decision stream controlling actions that emerge from the rate equations.

10.2 Model Structure and Behavior

10.2.1 Case Study

Potato-based starch was first introduced by Frenudco in 2009, establishing a small manufacturing unit of 1000 ton capacity near Sundar Industrial Estate. Frenudco extracts wet potato starch from potatoes and then convert it into dry potato starch while passing it through the drying unit. After getting 90% degree of dryness, dry starch is sent to Lahore for grinding purpose. Management is not willing to invest in grinding mill as the toll manufacturing is available at a reasonable price and relieves the management of the hassles of grinding and packaging. Potato chips manufacturers also generate wet potato starch as a by-product that further enhances the capacity of the firm annually. The firm's first priority is to take the extracted wet potato starch from potato chips manufacturers as an exogenous variable to simplify the model. Frenudco is bound to take wet potato starch at an average of around 90 tons per month as per the agreement. In reality, the supply of wet starch in peak months is more than 110 tons per month, but during the annual plant shutdown, wet starch production is zero. So on an average, Frenudco gets 90 tons of wet potato starch per month from the chips-producing companies. Downstream processes like drying, grinding, and packaging are performed to add value to the product. Management of Frenudco attempts to maintain a maximum inventory of grinded potato starch of 50 tons, limited by the capacity of finished goods warehouse. When the inventory of grinded potato starch falls below this level, workers in the grinding department start grinding and packaging processes. When the inventory of grinded starch grows considerably above this level, grinding and packaging operation is stopped to reduce the departmental expenditures. The management issues a directive to the sales team that finished potato starch shipment rate should not be higher than the plant capacity to avoid the delivery delays and backlog. The management would rather limit the growth than to earn the bad name due to high delivery delays and back ordering. The management would like to find out the effect of policies like workers contribution to drying department, recruitment of the new hires in drying section, and enhancement of the labor productivity that can turn the fate of the company. Also, there are some other avenues yet to be explored.

10.2.2 Push and Pull Production Decisions

Frenudco Manufacturing is a firm with many diverse operations aimed at meeting the various requirements of the product depending upon its usage. Extracting wet potato starch from potatoes, converting it into dry potato starch, and then grinding and packaging are the core processes that cannot be avoided during the production operations. Grinding and packaging are two distinct processes but are usually done simultaneously due to the product nature; the grinded material is immediately packed in a moisture-free polyethylene bags to maintain a specified degree of dryness. The firm involves key operational processes (Beamon, 1998) such as handling the wet potato starch having 54% degree of wetness, converting it into dry starch with up to 10% degree of wetness, and then grinding and packaging the starch to convert it into a required type of product depending upon its industrial usage using different mesh sizes from 200 and onward. Fluctuating work in process inventories (Croson and Donohue, 2005) in all three stages—wet potato starch, dry potato starch, and grinded potato starch—is the major issue faced by the firm. The storage capacities for wet starch and dry starch are 1000 and 500 tons, respectively, whereas the storage capacity for the grinding and packaging departments is limited to 50 tons. For drying and grinding department, production strategy is push based, build the stocks and make the product available on the pull of customers' request following pull-based production strategy (Goncalves et al., 2005).

To put wet starch into drying mode is highly labor intensive, and the overall production of the drying department is directly proportional to the number of workers assigned to the drying department. The capacity of the drying equipment, the motivational level of workers in the drying department, and the dry weather are the additional factors that enhance the drying rate. Labor productivity in the drying department, calculated on an average basis, is close to 15 tons per month per worker. To initiate the control measure, the firm has decided to have six workers in the drying department to make the wet starch dried at its earliest assuming the citrus peri bus. Dried potato starch is stored in 50 kg bags and then it is ready for next operations of grinding and packaging. Grinding and packaging process is performed almost 50 km away from the drying site due to some constraints of manufacturing unit. Grinding of the dried potato starch is done using different mesh sizes to control the particle size as per the product demand, then it is bagged using the inner polyethylene sheet to provide the moisture barrier, and finally, sealing operation is performed. The capacity of the grinding and packaging department is limited to 50 tons per month. The labor productivity of the workers involved in grinding and packaging department is 10 tons per month per worker. The management is ambitious to maintain the stock up to 50 ton over a period of 4 months (grinding adjustment time is 4 months). For this purpose, the management has approved the budget for three workers in the grinding and packaging department.

10.2.3 Sales Operation

Frenudco started his sales from a single customer whose projected estimated requirement was 60 tons per month, but due to operational teething problems, the requirement was restricted only to 5 tons per month. Over the next 15 months, the customer base increased from one customer to five customers, who were using corn-based starch since its inception and changed to potato-based starch partially due to its better mileage, low cost, and quick delivery. This raised the requirement from 5 to 30 tons per month. All customers of Frenudco are industrial clients, and their demand depends on their own consumption during the manufacturing process. Operational parameters such as shortage of gas, electricity load shedding, machinery breakdowns, and nonavailability of chemicals due to strikes, etc., are the key variables that affect the production rate of the starch consumers. This ultimately affects the sale of the Freundco starch. Every customer has different machinery and operational process parameters. Even the customers of the same industry have the machinery of varied degrees of automation and technology that hampers the sale rate of potato starch. Once the customers start buying potato starch, they remain loyal to it. The customer turnover is zero, so far. Potato starch is difficult to sell to industrial clients because each client will have to conduct the test run first in the laboratory and then on the production machine to fine-tune the operational process parameters, which vary from customer to customer. The customer's word-of-mouth contributes more to the sale of potato starch and helps to find a new customer than the salesman effectiveness or any other promotional incentive scheme. Customer induction rate is directly proportional to the stock of existing customers and the customer-finding efforts factor. Every 9 months, on average, each salesman signs up one new customer (0.1 customer-finding efforts factor). An average order size of a single customer is 5 tons per month. The customer's loyalty and stay with the company is nearly 10 years (120 months).

10.2.4 Employment Practices

Frenudco has had a very stable employment history. The company has maintained 12 employees who have been distributed traditionally; 6 employees to the drying department, 3 to the grinding and packaging department, and 3 to the sales department. The company has not fired anybody since its inception and is very careful while hiring new workers. The tasks performed by individuals are very specific in nature. The geographical displacement of different functional units also discourages multiskilling. On-the-job training is a contributing factor for capability development. Employee turnover is almost zero due to handsome salary packages, bonuses, and other fringe benefits like travel allowance, daily allowance, medical facility, free residence, and company-provided meals.

10.2.5 Feedback Structure

Model feedback notions are translated first into block diagrams and finally in the DYNAMO (dynamic models) machine-readable computer code (Alexander, 1980). Anderson et al. (2012) presented a deterministic, mixed-integer, nonlinear mathematical programming model, based on economic order quantity (EOQ) techniques. Here, the limiting factor is the capacity of the grinding and packaging warehouse.

The model is the sixth-order differential equation with associated flows (Forrester, 1968). The order of the model depends upon the number of levels and the number of delays (Sushil, 1993). As in this model there is no delay function, the number of levels determines the order of the model (Barlas and Yasarcan, 2005). The polarity of the positive and negative feedback loop generates behavioral patterns (Anderson and Sturis, 1988). The system consists of major positive and negative feedback loops.

10.2.5.1 Reinforcing Loops

Loop (1):

Customers → (+) Shipments → (-) Grinded Potato Starch → (-) Customers

Loop (2):

Grinded Potato Starch → (+) Shipments → (+) Customers → (+) Workers in Drying Dept → (+) Dry Potato Starch → (+) Grinded Potato Starch

Loop (3):

Customers → (+) Customer-Finding Rate → (+) Customers

Major positive feedback loop is apt to appear in its degenerative form. A higher stock of grinded potato starch exerts the pressure to speed up shipments and to conduct production trials at the customer end. Successful trials encourage the customer to accept the incoming supplies of potato starch. That is why the stock of grinded potato starch declines at the grinding and packaging stage. The empty warehouse of the grinded potato starch encourages the management to start the grinding and packaging process and fill up the capacity of the warehouse over a period of 4 months. Low customer order base does not exert the pressure on the workers in the grinding and packaging department. The workers of the grinding department are different from drying department, and interdepartment transfers are totally restricted. There are multiple factors involved in the drying process, and they make the process complex (Lyneis, 1980). Keeping wet potato starch in the production floor affects the product quality, and the starch loses its efficacy for the manufacturing of certain products. Ideally, there should be no wet potato starch on the production floor. But factors like high incoming rate of wet potato starch and weather conditions threaten the system's viability unless it is checked by some self-regulating balancing loop. Customers attract other customers. The success of the order executed in one customer spreads the word-of-mouth in the market that helps to find new customers.

In the major negative feedback loop, reduced inventory of the grinding department is compared with desired inventory limited to warehouse capacity. Discrepancy in inventory is a guide to action; first, for grinding and packaging operation and second for enhancing the drying effectiveness, thereby overcoming the drying issues. Productivity enhancement incentive is the key factor to build the labor productivity in the drying department.

If the negative loop dominates, it will negate the external disturbance, and if the polarity of influence is too small, then positive loop will dominate. Different behavior modes are understood with coherent pieces of structures and loop dominance (Oliva et al., 2006). In our case, the polarity of positive loop dominates and system collapses in the long run while building the stock of wet potato starch and backlog of grinded potato starch. A high backlog of grinded potato starch will reduce customers in the long run when delivery delays will be high. At the moment, the time horizon is up to 50 months and customers' requirements are equated to plant capacity.

Intentionally, the aspect of chaotic structures (Anderson and Sturis, 1988) in the model is neglected; confining the discussion to the policy depicts the unhealthy condition of the company. At the end, polices are proposed on the basis of parametric changes and structural changes to make the system better behaved (Williams, 2000).

Negative loops are goal seeking and adjust activity toward some target value.

The balancing loops are as follows:

Loop (4):

Workers in Grinding and Packaging Dept → (+) Grinded Potato Starch → (+) Discrepancy in Grinded Potato Starch → (–) Workers in Grinding and Packaging Dept

Loop (5):

Workers in Drying Dept → (+) Dry Potato Starch → (+) Discrepancy in Dry Potato Starch → (–) Workers in Drying Dept

Loop (6)

Customers → (+) Customer Leaving Rate → (-) Customers

Loop (7)

Workers in Drying Dept → (-) Workers transformation rate to Drying Dept → (+) Workers in Drying Dept

Base Run Graph 1 indicates the stock of wet potato starch and dry potato starch. Over a period of 50 months (almost 4 years), the stock of wet potato starch keeps on increasing until it reaches the maximum point, whereas the stock of dry potato starch reaches its maximum level in 25 months, indicating the effectiveness of the workers of the drying department. After 22 months, the stock of the drying department starts declining. There are two reasons for reduction in the stock levels of dry potato starch. First, the increase in customer orders; therefore, whatever quantity is dried, it is grinded and sold to customers. Second, the huge stocks of wet potato starch demotivate the limited staff of the drying department. That affects the labor productivity. Base Run Graph 2 shows the customer's trend, which is increasing and puts pressure on the management. The management has to line up their resources to cope up with the market demand; otherwise, the shipment rate of the finished starch will not be more than the current capacity of the plant. After 5 months, the stopped grinding process fraction will be more than zero, compelling the workers of the grinding and packaging department to have the stock of grinded starch to avoid stock outs. But it is quite obvious that after 5 months grinded stock would be out of stock to meet the customer's demand and whatever the quantity has been grinded is sold out immediately. Negative figures of grinded potato stock depict the backlog of orders. The Cartesian frame of dry potato starch and grinded potato starch indicates the interesting behavior of the grinded potato starch. Over the period of 50 months, there are some ups and downs in the values of grinded potato starch as well as dry potato starch (Figures 10.1 and 10.2).

10.3 Policy Experiments

Models created for policy design perspective must incorporate multiple patterns potentially existing in the system and observed and

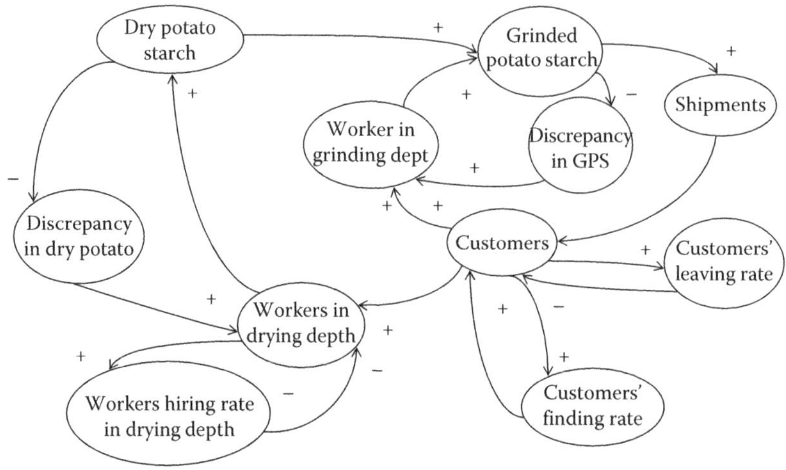

Figure 10.1 Potato starch model.

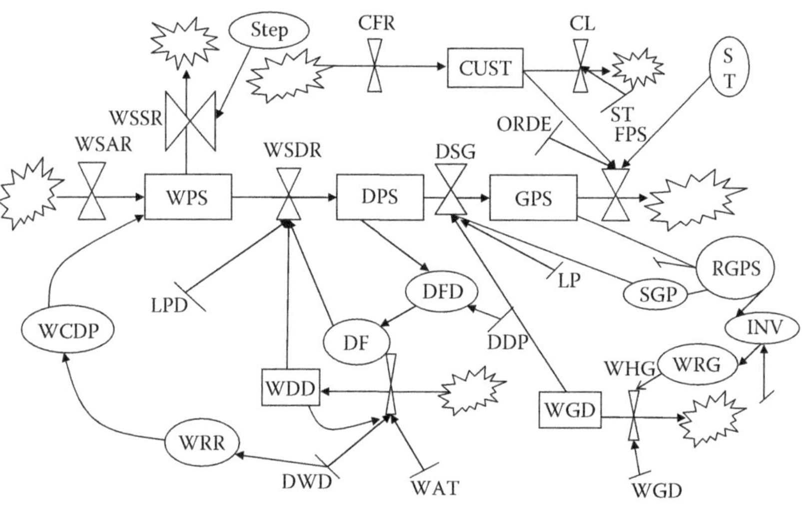

Figure 10.2 Stock and flow diagram.

recorded at different times and locations (Lewis and Baudians, 2007) so that the mechanisms of change from one pattern to another can be searched through experimentation (Saeed, 2008). Experimentation with the model not only allows us to understand the diversity of patterns but also helps us unveil the loops and critical elements that have pronounced effect on the rising stocks of inventories. Various policies were tested in the following simulations. Primarily, policies tested are based on sensitive parameters. For this type of scenario, it is assumed that the policy maker is satisfied with the equations, and the only doubt remains as to what values the parameters or the policy levers should have (Duggan, 2008). Second, policy experiments are performed on the structural changes having a base of mental intuition and perception about the problem solution. This is the equation perspective to optimization.

10.3.1 Model Response to Parametric Changes

10.3.1.1 Policy Run 1: Grinding Adjustment Time Programs Grinding adjustment time (GAT) is a sensitive parameter. Our routine practice was to grind potato starch in such a way that our stock should rise to desired level of 50 tons in 4 months. As shipments are very nominal most of the time—around 20 tons per month. Our pre-simulation prediction was that it is most appropriate and there is no need to change this grinding pattern. But when we reduced the GAT from 4 months to four times per month (0.25 month), efforts are made to grind the material after every week to build the stock in small replenishments instead of grinding in one go after every 4 months. This policy works well, and the stock of grinded potato starch is available all the time and there is no risk of stock out of finished potato starch generating any backlogs. The increased level of grinded potato starch does not give the confidence to customers that the firm has enough stock to meet the demand all the time but as well highlights the management's perception about the customer. The management is intestinally willing to bear the inventory-carrying cost but cannot afford to stock out in this competitive environment.

10.3.1.2 Policy Run 2: 50% Increase in the Staffing of Drying Department The rising stock of wet potato starch exerts pressure on the management to increase the number of workers in the drying department because drying effectiveness has a direct link to the number of workers in the drying department. It is suggested that the company have a 50% increase in the staffing of the drying department and raise the number of workers in the drying department from six to nine. Our pre-simulation prediction was that our drying effectiveness would significantly increase; the effect of this is observed in the initial months when all wet starch is converted into dry potato starch. Negative values of the wet potato starch shows that the workers of the drying process are underutilized while they are capable of converting higher quantities of wet starch into dry potato starch.

10.3.2 Model Response to Structural Changes

10.3.2.1 Policy Run 3: Worker Contribution in Drying Process: Productivity Improvement Program Workers' contribution during the drying process enhances the drying rate. Consequently, the stock of dry potato starch increases. The maximum value that we get over a period of 25 months is 600 tons. Later on, workers' contribution starts decreasing and end ups with a stock of 102 tons. Workers' contribution is perceived and mapped in the model through the graphical function TABHL, taking the ratio of workers required, on the x-axis, to their contribution in drying process, on the y-axis. The workers required ratio is between available workers in the drying department and desired workers in the drying department. If the number of available workers in the drying department is closer to desired workers, then team synergy builds up in the department. As a result, contribution toward the drying process is enhanced.

10.3.2.2 Policy Run 4: Selling Wet Starch at Reduced Rate The rising stocks of wet potato starch in the base run exert pressure on the management to reduce the stocks of wet potato starch. The efforts made to increase the drying rate enhance the productivity of workers in the drying department. It is also questioned by using the test function that if we start selling the wet potato starch at reduced rates in order to reduce the stock of potato starch, shall it work? After 10 months when the stock of wet starch becomes difficult to manage, the management implements this policy, and results are quite encouraging. This policy helps to reduce the stock level and thus minimize the financial loss, which is linked with the purchase of wet potato starch. The financial sector discussion is out of scope of our model. The selling quantity of wet potato starch is not more than 15 tons.

10.3.2.3 Policy Run 5: Technological Deployment for Dryness Factor The management has specified the space in the drying department to store the stock of 1000 tons of the potato starch. It helps to reduce the stock of wet potato starch and also provides the cushioning effect on the grinding and packaging department to make the material available for grinding purpose. To enhance the dryness factor, technological modifications are required to ensure the availability of Sui gas for the drying process. The policy of biogas installation and modification in the machinery is proposed to enhance the dryness of wet starch and reduce the discrepancy between desired dry potato starch and actual available dry potato starch. But simulation results are not very encouraging; along with the combination of other policies, but alone, the result of this policy is more or less same.

10.4 Conclusion

Policy experiments indicate that interactions between different functions of the organization (Akhermans and Dellaert, 2005) and market are so complex that they cannot be intuitively appreciated. A single policy may affect the performance of one department, but the overall impact of that policy is not significant. But when we combine all policies, the results give the insight into the problem under study. The concept of feedback system structure for organizing the explicit and tacit knowledge about the system (Saeed, 2008) explores the avenues that make the system better behaved over a period of time. The translation of dynamic hypothesis into a model provides insight, to the management, into the causes of behavior. This chapter formulates the policies for the management of Frenudco. The policies are thought-provoking and eye-opening because pre-simulation predictions of the management were contrary to the actual policies suggested by the model. These policies address the issues of huge inventories, indicate the areas for the effective utilization of resources, and highlight the budgeting and investment opportunities.

Appendix 10.A: Tables

See Tables 10.A.1 and 10.A.2.

Table 10.A.1 Symbols for Flow Diagram

DESCRIPTION	SYMBOL	ASSOCIATED EQUATION TYPE	EXPLANATION
Level		L	Stock
Rate		R	Flow
Auxiliary		A	Convertor
Table function		T	Perception map between x and y plane
Exogenous variable		E	Occasionally affect the model behavior but not part of model
Constant		C	Constant that has unique value
Source or sink of material		Define	Out of boundary, defines the model scope
Material/ information flow			Use for the movement of material and information

Table 10.A.2 Products of Potato Starch

PRODUCT	USAGE	INDUSTRY
A	For animal feed manufactures feedmill	Livestock
B	Sizing	Textile
C	Glue manufacturing	Packaging mills
D	Paper sizing	Paper and board mills
E	Oxydry powder to control offsetting	Printing
F	Liquid glue	Core making book binding
G	MIAIA	Cloth

Appendix 10.B: List of Variables

VARIABLES	DESCRIPTION	UOM	EQUATION TYPE	PARAMETRIC VALUE
WPS	Wet potato starch	Tons	L	0
DPS	Dry potato starch	Tons	L	0
GPS	Grinded potato starch	Tons	L	0
WDD	Workers in drying department	No. of workers	L	0
IWDD	Initial workers in drying department	No. of workers	C	6
WGP	Workers in grinding and packaging department	No. of workers	L	0
CUST	Customers	No. of customers	L	0
ICUST	Initial customers	No. of customers	C	1
IWPS	Initial wet potato starch	Tons	N	0
IDPS	Initial dry potato starch	Tons	N	0
IGPS	Initial grinded potato starch	Tons	N	0
WSAR	Wet starch arrival rate	Tons per month	R	
WSDR	Wet starch drying rate	Tons per month	R	
WSSR	Wet Starch Shipment Rate	Tons per month	R	
LPD	Labor productivity in drying department	Tons per month per worker	C	15
LPG	Labor productivity in grinding and packaging department	Tons per month per worker	C	10
DDPS	Desired dry potato starch	Tons	C	1000
DFDDPS	Difference in desired dry potato starch	Tons	A	
DF	Drying factor	Dimensionless	A	
TDF	Table for drying factor	Dimensionless	T	
DSGR	Dry starch grinding rate	Tons per month	R	
FPSR	Finished product shipment rate	Tons per month	R	
SGP	Stop grinding process	Dimensionless	A	
TSGP	Table function for stopping the grinding process	Dimensionless	T	
WCDP	Workers' contribution in drying process	Dimensionless	T	
TWCDP	Table for workers' contribution in drying process	Dimensionless	T	
WRR	Workers required ratio	Dimensionless	A	
RGPS	Ratio	Dimensionless	A	
AFACT	Average wet potato starch from factory	Tons per month	A	
EXTFACT	Wet potato starch extracted from factory	Tons per month	C	90

(*Continued*)

VARIABLES	DESCRIPTION	UOM	EQUATION TYPE	PARAMETRIC VALUE
DISGPS	Discrepancy in grinded potato starch	Tons per month	A	
WBG	Workers budgeted in grinding department	No. of workers	A	
CAP	Production capacity	Tons	C	90
WRGD	Workers required in grinding department	No. of workers	A	
INVC	Inventory correction	Tons	A	
IAT	Inventory adjustment time	Months	C	1
WHDD	Workers hired in drying department	No. of workers per month	R	
WATD	Workers adjustment time in drying department	Months	C	1
WHGP	Workers hired in grinding and packaging department	No. of workers per month	R	
WATG	Workers adjustment time in grinding and packaging department	Months	C	1
DWGDP	Desired workers in grinding and packaging department	No. of workers	C	
DWDD	Desired workers in drying department	No. of workers	C	6
CFR	Customer finding rate	No. of customer per month	R	
CLR	Customer leaving rate	No. of customer per month	R	
CRFM	Customer finding coefficient	No. of customer per month	C	0.1
SLC	Stay length with the company	Months	C	120
GAT	Grinding adjustment time	Months	C	4
DGPS	Desired grinding potato starch	Months	C	50
ORDER	Order size per customer	Tons	C	5
WSSR	Wet starch shipment rate	Tons per month	R	
STEP	Step function	Tons	A	
RAMP	Ramp function	Tons	A	
WSQTY	Wet shipment	Tons	C	10
TESTWS	Test for wet shipment	Tons	A	
POLY1	Policy 1		A	
POLY2	Policy 2		A	
POLY3	Policy 3		A	
POLY4	Policy 4		A	
POLY5	Policy 5		A	

(*Continued*)

VARIABLES	DESCRIPTION	UOM	EQUATION TYPE	PARAMETRIC VALUE
SW1	Switch function			1
SW2	Switch function			0
SW3	Switch function			0
SW4	Switch function			0
SW5	Switch function			0
SW6	Switch function			0

Appendix 10.C: Variables with Base Run and Policy Run Parametric Values

VARIABLES	DESCRIPTION	UOM	BASE RUN PARAMETRIC VALUE	POLICY RUN PARAMETRIC VALUE
IWDD	Initial workers in drying department	No. of workers	6	6
ICUST	Initial customers	No. of customers	1	1
LPD	Labor productivity in drying department	Tons per month per worker	15	18
LPG	Labor productivity in grinding and packaging department	Tons per month per worker	10	10
DDPS	Desired dry potato starch	Tons	1000	1000
EXTFACT	Wet potato starch extracted from factory	Tons per month	90	90
CAP	Production capacity	Tons	90	90
IAT	Inventory adjustment time	Months	1	1
WATD	Workers adjustment time in drying department	Months	1	1
WATG	Workers adjustment time in grinding and packaging department	Months	1	1
DWDD	Desired workers in drying department	No. of workers	6	6
CRFM	Customer finding coefficient	No. of customer per month	0.1	0.1
SLC	Stay length with the company	Months	120	120
GAT	Grinding adjustment time	Months	4	0.25
DGPS	Desired grinding potato starch	Months	50	50
ORDER	Order size per customer	Tons	5	5
WSQTY	Wet starch shipment quantity	Tons	10	10

Appendix 10.D: Program for System Dynamics Simulation Model on DYNAMO Software

```
Note                    Potato Starch Model
Note ********************************************
Note                     Wet Potato Starch
Note ********************************************
L WPS.K=WPS.J+DT*(WSAR.JK-WSDR.JK-WSSR.JK)
N WPS=IWPS
C IWPS=0
R WSAR.KL=AFACT.K
A AFACT.K=EXTFACT
C EXTFACT=90
R WSDR.KL=WDD.K*LPD.K*DF.K*WCDP.K
C LPD.K=POLY4.K
Note *********************************************
Note                     Dry Potato Starch
Note *********************************************
L DPS.K=DPS.J+DT*(WSDR.JK-DSGR.JK)
N DPS=IDPS
C IDPS=0
R DSGR.KL=((WBG.K*LPG)+(DISGPS.K/GAT.K))*SGP.K
C LPG=10
A WBG.K=MIN(3,WGPD.K)
A GAT.K=POLY3.K
C DDPS=1000
A DFDDPS.K=DDPS-DPS.K
A DF.K=TABHL(TDF, DFDDPS.K,0,1000,100)
T TDF=0/0.05/0.15/0.25/0.35/0.45/0.55/0.65/0.7/0.8/1
Note *********************************************
Note                   Grinded Potato Starch
Note *********************************************
L GPS.K=GPS.J+DT*(DSGR.JK-FPSR.JK)
N GPS=IGPS
C IGPS=0
A RGPS.K=PRLF.K/DGPS
A PRLF.K=MIN(50,GPS.K)
A SGP.K=TABHL(TSGP,RGPS.K,0,1,.1)
T TSGP=1/0.93/0.75/0.65/0.5/0.35/0.35/0.2/0.15/0.1/0
A DISGPS.K=DGPS-GPS.K
A INVC.K=DISGPS.K/IAT
C IAT=1
C DGPS=50
A WRGD.K=INVC.K/LPG
```

```
R FPSR.KL=MIN ((CUST.K*ORDER*TEST1.K),CAP)
C CAP=90
A TEST1.K=SWITCH(1, (1+STEP0.1,5)),SW1)
C SW1=1
Note ***********************************************
Note                Workers in Drying Department
Note ***********************************************
L WDD.K=WDD.J+DT*(WHDD.JK)
N WDD=IWDD
C IWDD=6
A DWDD.K=POLY5.K
R WHDD.KL=(DWDD.-WDD.K)/WATD
C WATD=1
A WCDP.K=TABHL(TWCDP, WRR.K,0,1,0.25)
T TWCDP=0/0.5/0.7/0.85/1.1
A WRR.K=WDD.K/DWDD.K
Note ***********************************************
Note     Workers in Grinding and Packaging Department
Note ***********************************************
L WGPD.K=WGPD.J+DT*(WHGD.JK)
N WGPD=IWGPD
C IWGPD= 0
R WHGD.KL=WRGD.KWATG
C WATG=1
Note ***********************************************
Note                           Customers
Note **********************************************
L CUST.K=CUST.J+DT*(CFR.JK-CLR.JK)
N CUST=ICUST
C ICUST=1
R CFR.KL= CUST.K*CFRM
C CFRM=0.1
R CLR.KL= CUST.K/STC
C STL=120
C ORDER=5
Note ***********************************************
Note                    Alternative Policies
Note ***********************************************
A POLY1.K=SWITCH(1,WCDP.K,SW2)
C SW2=0
A POLY2.K=SWITCH(0,TESTWS.K,SW3)
C SW3=0
R WSSR.KL=WSQTY*POLY2.K
```

```
C WSQTY=10
A TESTWS.K=1+STEP(0.5,10)
A POLY3.K=SWITCH(4,0.25,SW4)
C SW4=0
A POLY4.K=SWITCH(15,18,SW5)
C SW5=0
A POLY5.K=SWITCH(6,9,SW6)
C SW6=0

Note ************************************************
SPEC              DT=0.125/SAVPER=1/LENGTH=50
Note ***************************************T********
SAVE WPS,DPS,GPS,WDD, WGPD,CUST
SAVE WSAR,WSDR,DSGR,FPSR,CFR,CLR
SAVE WSSR, WHDD,WHGD,INVC,SGP,WCDP,DF,
SAVE POLY1,POLY2,POLY3,POLY4,POLY5
```

Appendix 10.E: Behavior Patterns

Base Run Graph 1: Wet potato starch and dry potato starch.

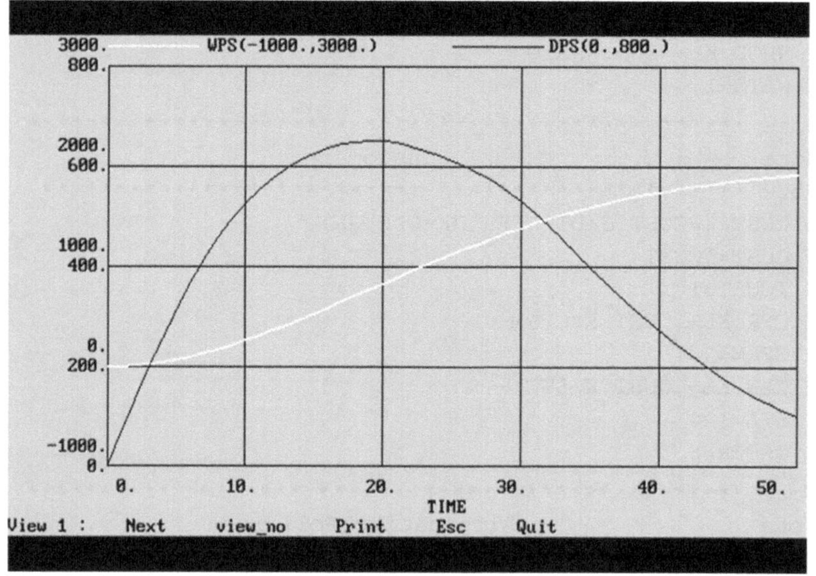

Base Run Graph 2: Grinded potato starch and customers.

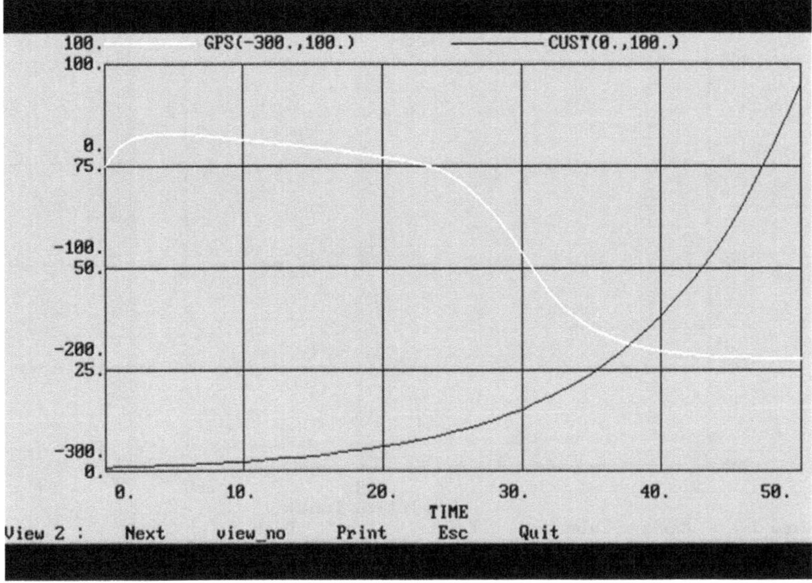

Base Run Graph 3: Wet potato starch and dry potato starch.

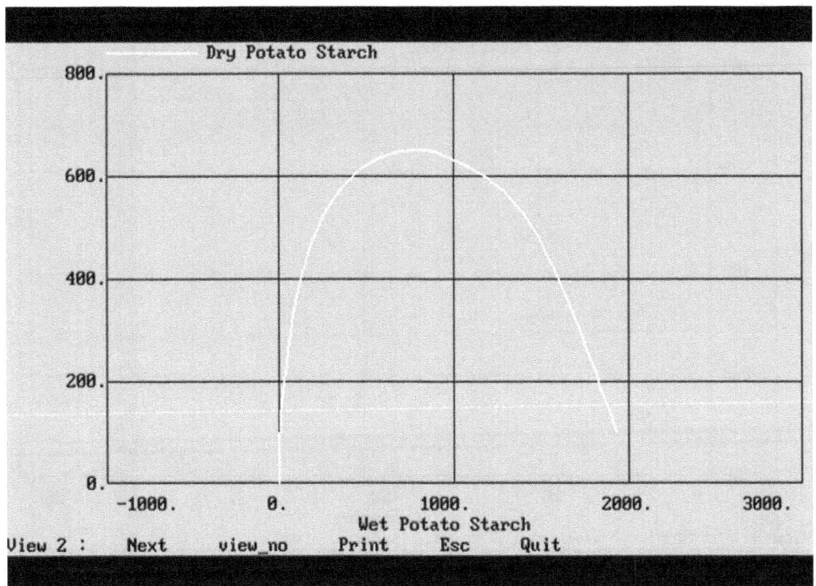

Base Run Graph 4: Dry potato starch and grinded potato starch.

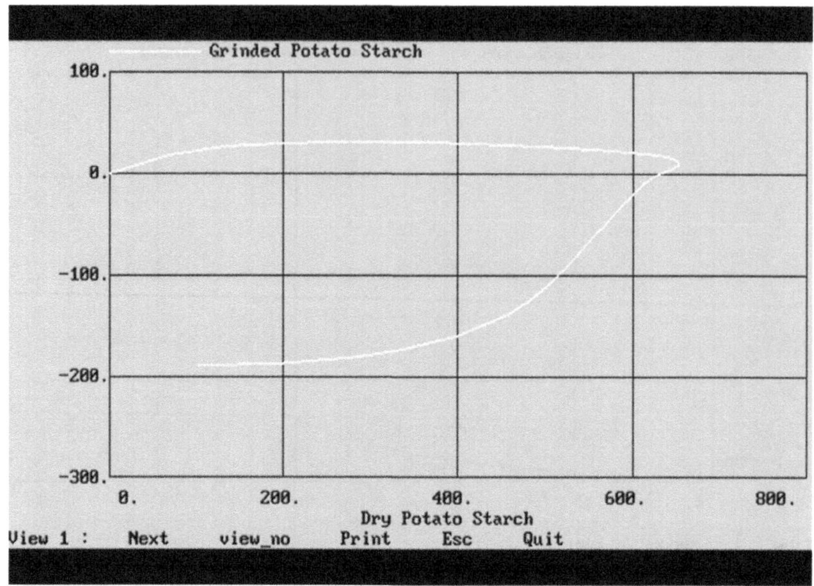

Base Run Graph 5: Customers and grinded potato starch.

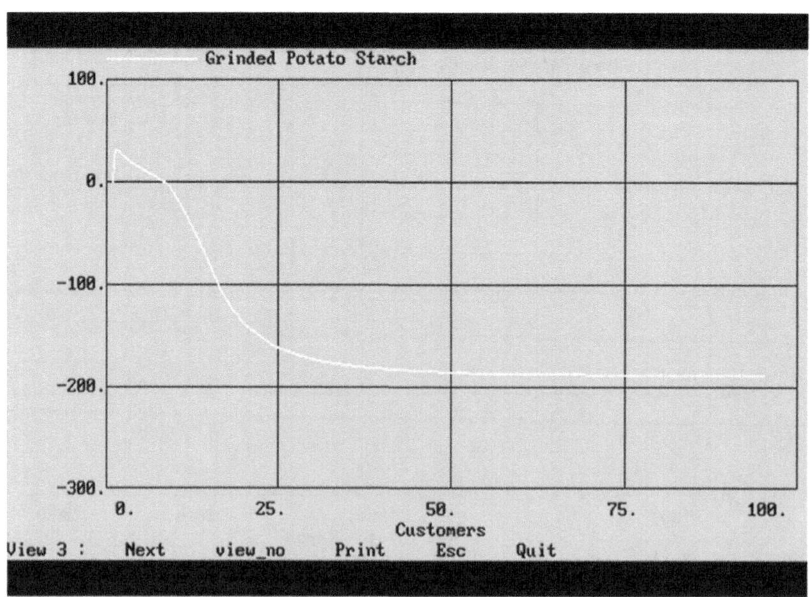

Policy run combining all policies—Graph 6.

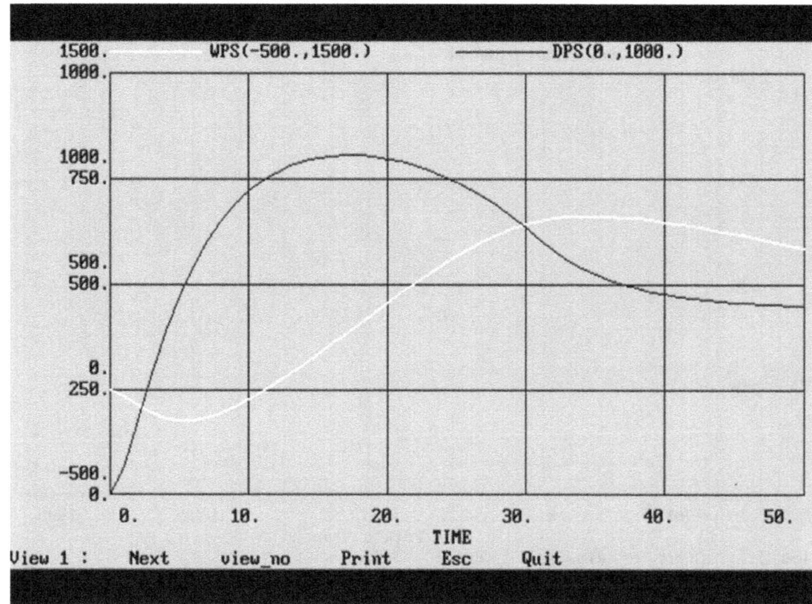

Policy run combining all policies—Graph 7.

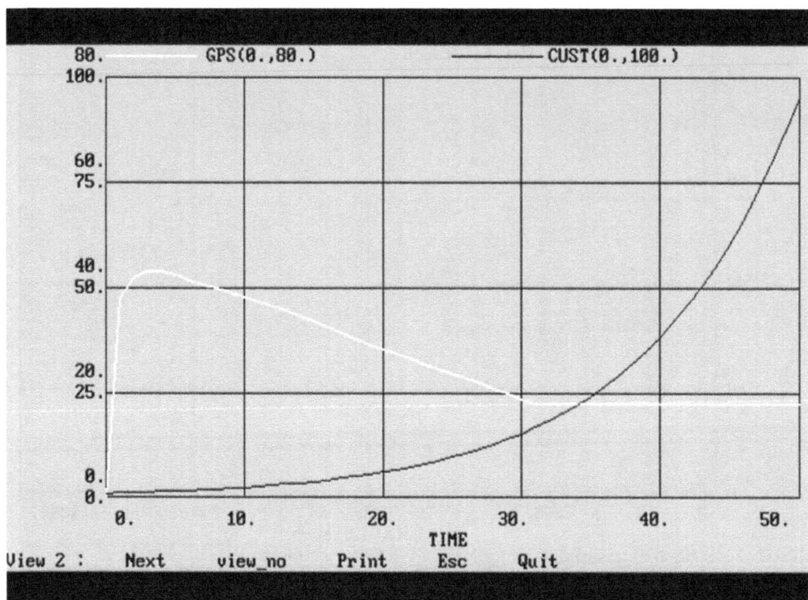

Policy run combining all policies—Graph 8.

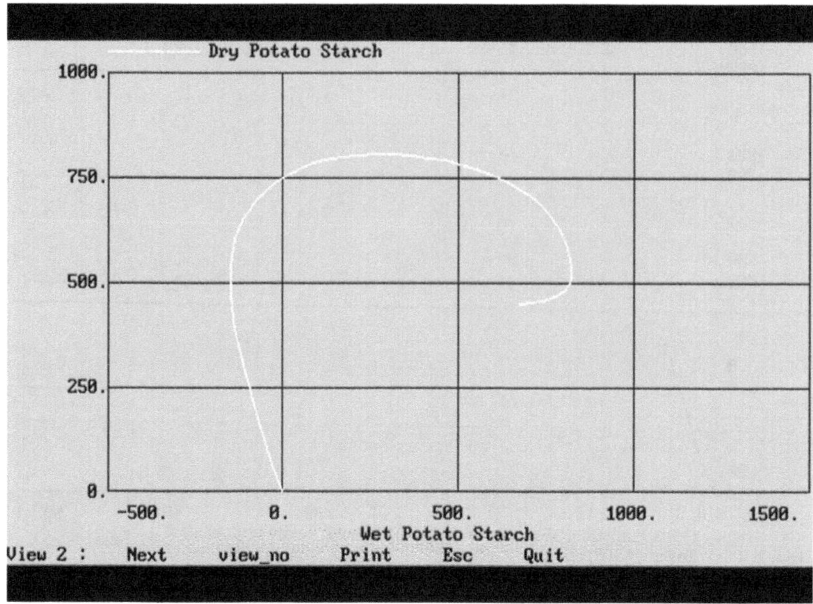

Policy run combining all policies—Graph 9.

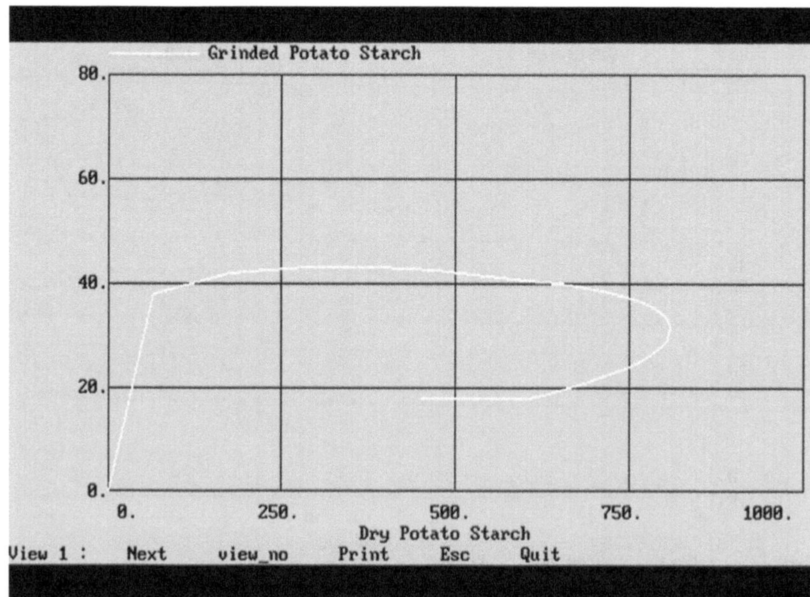

Policy run combining all policies—Graph 10.

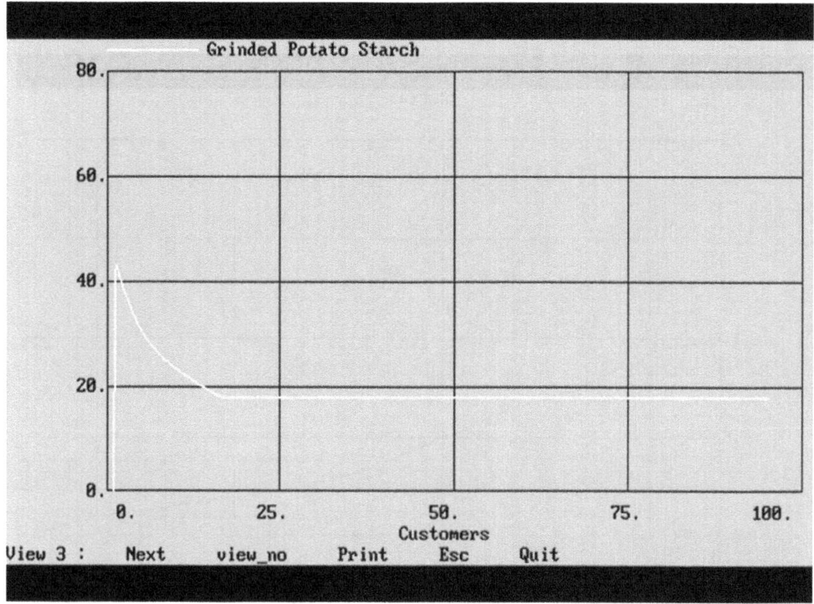

Table function (X = discrepancy in dry potato starch inventory, Y = dryness factor).

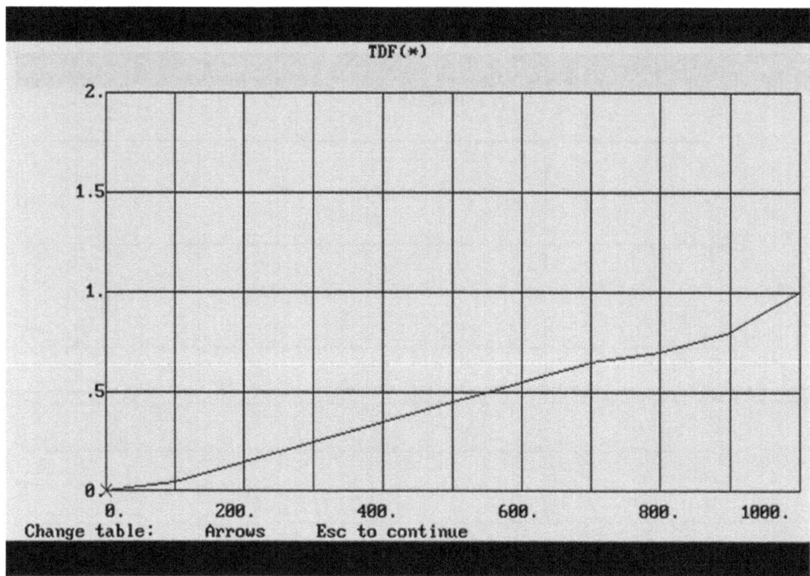

Table function (X = discrepancy in grinded potato starch inventory, Y = stop grinding process).

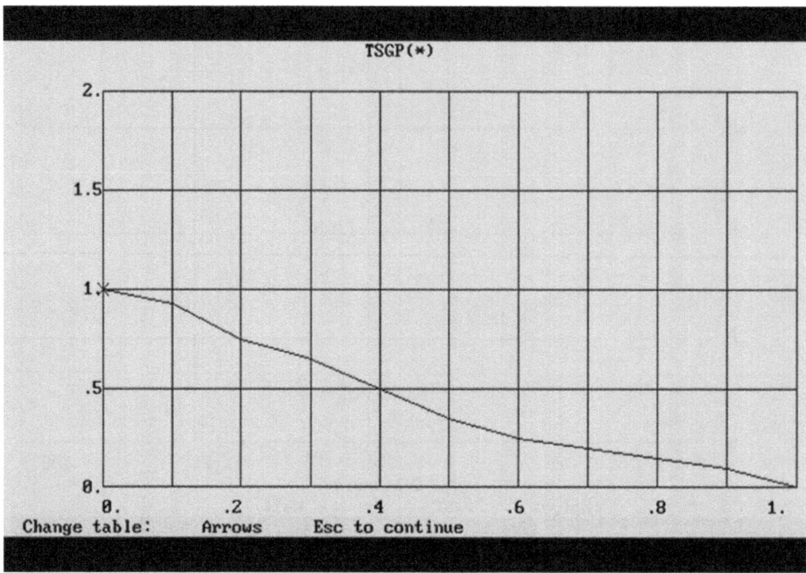

Table function (X = WDD/DWDD, Y = worker contribution in drying process).

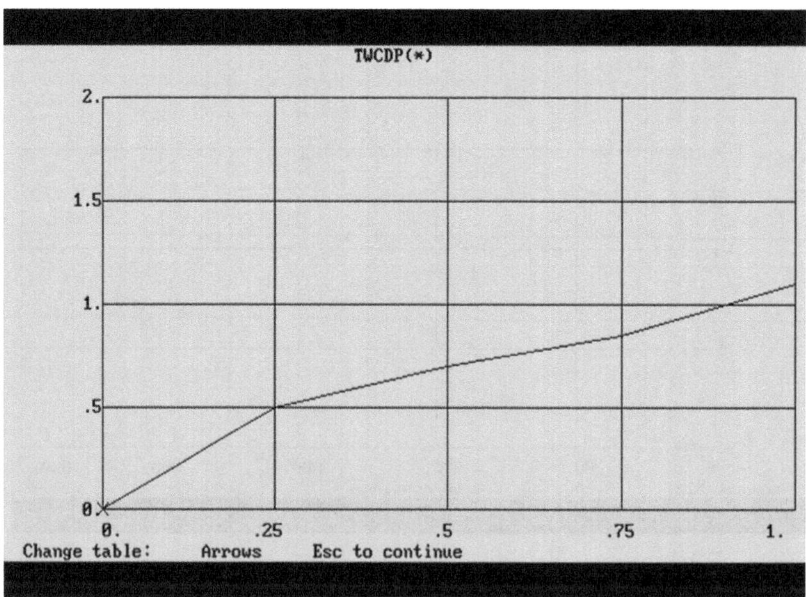

References

Akhermans, H. and Dellaert, N. (2005). The rediscovery of industrial dynamics: The contribution of system dynamics to supply chain management in a dynamic and fragmented world. *System Dynamics Review* 21 (3, Fall).

Alexander III, L.P. (1980). *Dynamo Manual.* MIT Press, Cambridge, MA, 1983.

Anderson, D.F. and Sturis, J. (1988). Chaotic structures in generic management model: Pedagogical principles and examples. *System Dynamics Review* 4 (1–2).

Anderson, D.R. et al. (2012). *An Introduction to Management Science Quantitative Approaches to Decision Making*. South-Western Cengage Learning Library of Congress, Mason, OH.

Barlas, Y. and Yasarcan, H. (2005). A generalized stock control formulation for stock management problems involving composite delay and secondary stock. *System Dynamics Review* 21 (1, Spring), 33–68.

Beamon, B.M. (1998). Supply chain design and analysis: Models and methods. *International Journal of Production Economics* 55 (3), 281–294.

Croson, R. and Donohue, K. (2005). Upstream versus downstream information and its impact on the bullwhip effect. *System Dynamics Review* 21 (3, Fall).

Duggan, J. (2008). Equation based policy optimization for agent-oriented system dynamics models. *System Dynamics Review* 24 (1, Spring), 97–118.

Forrester, J.W. (1968). *Principles of Systems.* MIT Press, Cambridge, MA.

Forrester, J.W. (1980). *Industrial Dynamics.* MIT Press, Cambridge, MA.

Goncalves, P. et al. (2005). The impact of endogenous demand on push-pull production system. *System Dynamics Review* 21 (3, Fall).

Howich, S. et al. (2006). Linking event thinking with structural thinking: Methods to improve client value in projects. *System Dynamics Review* 22 (2, Summer), 113–140.

Lane, C.D. (2007). The power of the bond between cause and effect: Jay Wright Forrester and the field of system dynamics. *System Dynamics Review* 23 (2–3, Summer/Fall).

Lewis, E. and Baudians, C. (June 2007). Whole systems thinking: Education for sustainability at a Montessori school. *ENGANA: Journal of Victorian Association for Environmental Education* 30 (1).

Lyneis, J.M. (1980). *Corporate Planning & Policy Design: A System Dynamics Approach.* MIT Press, Cambridge, MA.

Meadows, D.L. et al. (1994). *Dynamics of Growth in Finite World.* MIT Press, Cambridge, MA, 1974.

Oliva, R. et al. (2006). Loop eigenvalue elasticity analysis: Three case studies. *System Dynamics Review* 22 (2, Summer), 141–162.

Rabia, S. and Yusuf, I. (2007). A SD approach on quality education in class room environment of management schools. *Proceedings of the International Conference on System Science, Management Science and System Dynamics*, Shangai, China.

Richardson, G.P. (1986). Problems with causal-loop diagrams. *System Dynamics Review* 2 (2, Summer).

Richardson, G.P. and Pugh, A. (1981). *Introduction to System Dynamics Modeling with Dynamo*. MIT Press, Cambridge, MA.

Richmond, B. (1993). Systems thinking: Critical thinking skills for the 1990 and beyond. *Systems Dynamics Review* 9 (2, Summer).

Saeed, K. (2008). *Towards Sustainable Development: Essays on System Dynamics of National Policy*. Progressive Publishers, Lahore, Pakistan.

Sterman, J.D. (2007). Exploring the next great frontiers: System dynamics at fifty. *System Dynamics Review* 23 (2–3, Summer/Fall).

Sushil (1993). *System Dynamics a Practical Approach for Managerial Problems*. H.S. Popali for Wiley Eastern Ltd., New Delhi, India.

Towill, D.R. (1996). Industrial dynamics modelling of supply chains. *International Journal of Physical Distribution & Logistics Management* 26 (2), 174–196.

Weil, H.B. (2007). Application of system dynamics to corporate strategy: An evolution of issues and framework. *System Dynamics Review* 23 (2–3, Summer/Fall).

Williams, L. (2000). Designing instructional units with a systems thinking/system dynamics approach. *The Exchange* 9 (1, Winter), 1–12.

Yusuf, I. and Ramish, A. (2011). Modelling the factory dynamics: The case of CCRS. *International Conference on Business*, UCP, Lahore, Pakistan.

Author Index

V

W

Y

Z

Subject Index

T